THE CANAL BUILDERS

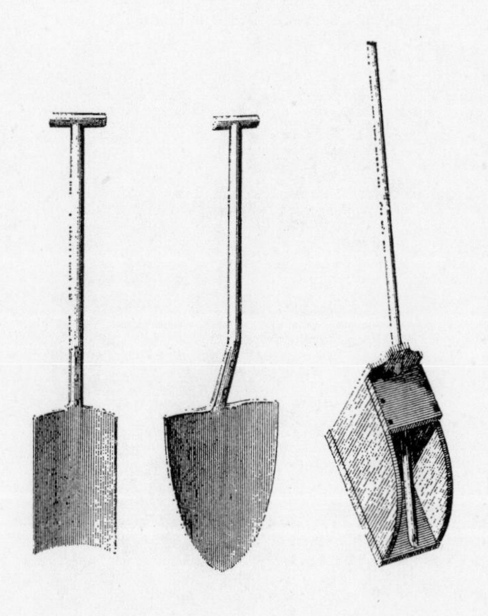

by the same author

A PROGRAMMED GUIDE TO OFFICE WARFARE

THE JONES REPORT

THE CANAL BUILDERS

ANTHONY BURTON

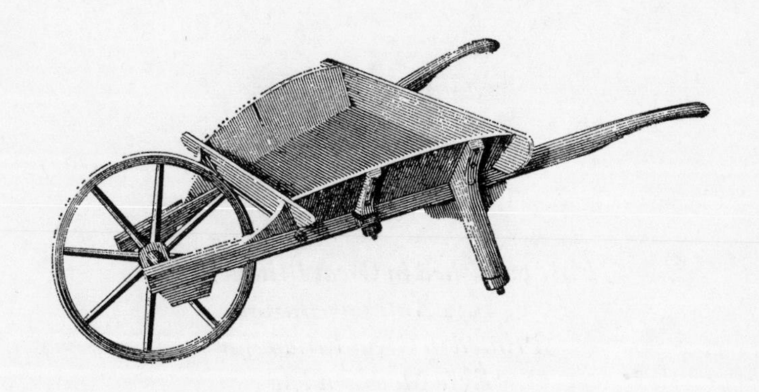

LONDON
EYRE METHUEN
1972

FOR JONATHAN, JENNY
AND NICHOLAS

First published in Great Britain 1972

Printed in Great Britain for
Eyre Methuen Ltd
11 New Fetter Lane, EC4P 4EE
by Cox & Wyman Ltd
Fakenham, Norfolk

SBN 413 28410 7

Distributed in the USA by
HARPER & ROW PUBLISHERS, INC.
BARNES & NOBLE IMPORT DIVISION

CONTENTS

ILLUSTRATIONS

DRAWINGS

MAPS

Acknowledgements and thanks for permission to reproduce photographs is due to Hugh McKnight for plates 1, 5, 8, 9, 10, 11, 25, 26, drawing 2 and the illustrations on the title pages and p. 218; to Josiah Wedgwood & Sons Ltd for plates 2 and 3; to the British Museum for plates 4, 6, 7, 12, 27, and drawings 1 and 3; to the National Portrait Gallery for plates 14, 16, 17 and 19; to the Institution of Civil Engineers for plates 15, 18 and 21; to the Waterways Museum, Stoke Bruerne, for plates 20, 22 and 29; to Sir Arthur Elton for plate 28; and to the Museum of British Transport for plates 30 and 31.
Map 1 is reproduced by kind permission of the British Museum. The rest of the maps were drawn by John Flower.

PREFACE

Anyone writing canal history today owes a tremendous debt of gratitude to the historians who have done the pioneering work in this field. In particular, I should like to acknowledge my debt to Charles Hadfield and L. T. C. Rolt, whose books and articles were the starting point for my own work. Many people have helped me directly in my search and by their comments and criticism. I especially offer my thanks to Mr E. H. Fowkes and the staff of the British Transport Historical Records Office both for giving me permission to use the material in the archives and for their help and encouragement.

January 1971, London

PREFACE

PART ONE

PROMOTERS AND FINANCIERS

1

THE BEGINNING OF THE CANALS

> Mr John Gilbert produced a Plan of a Canal or Cut proposed to be made from a Place in the Township of Salford to or near Worsley Mill . . . he is of opinion it is very practicable to make a Canal sufficient for the Navigation of Boats and Vessels of considerable Burthens, whereby the Carriage of Goods will be greatly facilitated and become less expensive.
>
> *Journal of the House of Commons*, 6 December 1758.

Mr John Gilbert was the agent of the Duke of Bridgewater. The canal plans that he described to Parliament were those for the Bridgewater canal – an undertaking which proved to be entirely practicable and every bit as efficient at improving transport as its proposers suggested. The first English canal to take a line across country which was independent of any natural waterway, it was the wonder of the day, and from its opening in 1761 became an essential stopping-off place for the eighteenth-century tourist.

> 'Tis not long since I viewed the artificial curiosities of London, and now have seen the natural wonders of the Peak; but none of them have given me so much pleasure as I now receive in surveying the Duke of Bridgewater's navigation in this county. His projector, the ingenious Mr Brindley, has indeed made such improvements in this way, as are truly astonishing. At Barton bridge he has erected a navigable canal in the air;

> for it is as high as the tops of trees. Whilst I was surveying it with a mixture of wonder and delight, four barges passed me in the space of about three minutes, two of them being chained together, and dragged by two horses, who went on the terras of the canal, whereon, I must own, I durst hardly venture to walk, as I almost trembled to behold the large river Irwell underneath me, across which this navigation is carried by a bridge, which contains upon it the canal of water, with the barges in it, drawn by horses, which walk upon the battlements of this extraordinary bridge.[1]

But if, for the tourist, the canal was no more than an object of curiosity and a titillation for refined sensibilities, for others it was the inspiration for an outburst of engineering activity such as Britain had not seen since the time of the Romans. Within half a century of the opening of the Bridgewater canal, the canal builders had covered Britain with an intricate network of artificial waterways. They revolutionized the transport system of the country.

The idea of a revolution in civil engineering sits uneasily with many of the images that we have of England in the middle of the eighteenth century. But the chronicles for the year 1761 show a time of contrasts: a pause between an older Britain and a new. The aspect of the age that is the best known and the most chronicled is the world of the wealthy and the fashionable, the London of the 1760s that Boswell described in his *Journal.* It was the age of reason, of Hume and Voltaire, and the age of sentiment and elegance. Laurence Sterne's verbal comic-strip of a novel, *Tristram Shandy,* was just starting to appear, Garrick ruled over the London theatre, the Adams and Chippendale were transforming the fashionable house and its furnishings, while Capability Brown was transforming the grounds outside. This fashionable world can be seen staring out at us from the paintings of Gainsborough and Reynolds. With the third George now on the throne, it all seemed very safe and stable. True, the Seven Years War was still stumbling along and in the course of the year 1761

[1] Anon., *A History of Inland Navigations* (1779).

the elder Pitt resigned from office, but these were not major disturbances. If power shifted between the Whigs and the Tories, it was still the same narrow social stratum that would provide the rulers. Electioneering, after all, was an occupation reserved for the very rich: at a 'small borough' the bill of fare for the electorate on polling day was recorded as '980 stone of beef, 315 dozen bottles of wine, 72 pipes of ale, and 365 gallons of spirits converted into punch'.[1]

But there was much in the year 1761 that looked back to an earlier time. It was still only fifteen years since the 'Bonnie Prince' had marched his highlanders down into England and only fourteen years since the Rebellion had been quelled at Culloden. Dipping into the chronicles, the reader keeps meeting items that present an older, and perhaps more robust, image of the times. In 1761, a sophisticated and respectable publication like the *Annual Register* could still find room in its pages for an anecdote about a rich Spanish visitor to London losing his trousers in a brothel, and the *Gentleman's Magazine* reported in April that 'an old sorcerer of 30 years standing was convicted at the quarter sessions of Norwich for defrauding a poor woman of money, by pretending to lay evil spirits and cure her of witchcraft'; and a letter from London on 12 May complained of 'the pernicious practice of driving cattle thro' the street of this city'. There are endless stories of highwaymen, some of whom actually seem to have lived up to their fictitious counterparts. When the dashing Isaac Dumas was acquitted, a poem was published by 'Certain Belles':

> Joy to thee, lovely thief! that thou
> Hast 'scap'd the fatal string;
> Let gallows groan with ugly rogues,
> Dumas must never swing.

He did swing later though. The lovely thief was hanged in the year 1761.

Behind the fashionable world, there was the constant discontent of the poor, which frequently burst out into riot. At Hexham,

[1] *Annual Register*, 1761, p. 101.

in Northumberland, the coal-workers rioted and the militia opened fire:

> The commanding officer ordered his men to fire over the heads of the rioters, but they, exasperated by the death of one of their officers, and two of their fellow militia-men, when once they began, were not to be kept within bounds. Think what a shocking sound! for near ten minutes, fellow subjects firing one upon another! and what a horrible scene did I behold afterwards, some carried by dead in carts, others on horses; and many were led along just dying of their wounds, and covered with blood! and to hear the dreadful shrieks of the women, whose husbands or sons were among the rioters, was enough to piece a heart of stone.... They reckon in all above 100 killed and wounded.[1]

There was discontent in the country as well as in the towns: 'A dispute having happened between the farmers of Kings-Langley and the Irish reapers, about wages, the latter, in order to oblige the farmers to comply with their demands, assembled, to the number of 200, armed with pistols, swords, guns and clubs, and threatened to fire the town.'[2]

Although it may not have preoccupied contemporary chroniclers, the most important factor of life in 1761 seems to us, looking back on it now, to be the continuing growth of industrial power. Already the great iron founders like the Darbys of Coalbrookdale were producing more iron and of better quality than ever before. Mining was being improved by the use of atmospheric engines for draining. New machines to improve manufacture, particularly in the textile industries, were being designed and developed. Only transport lagged behind among the innovations. Transport in 1761 was still in a wretched state and unless it could be improved and made cheaper, trade would be inhibited. Most important of all, unless transport could be improved the new towns of the Industrial Revolution would be starved of their most basic and essential commodity – coal.

[1] *Annual Register*, 1761, p. 83.
[2] *Gentleman's Magazine*, 1761, p. 378.

Land transport in the middle of the eighteenth century was little better than it was in the middle of the seventeenth. Goods were still carried by pack-horse trains along winding country tracks. Heavier and bulkier goods had to be taken in wagons and carts that churned the already muddy roads into a quagmire. Various attempts were made in Parliament to legislate better land transport into existence by setting limits to the loads that could be carried, and by stipulating the use of broad wheels. The efforts met with little success:

> The utility of broad-wheels, in amending and preserving the roads, has been so long and universally acknowledged, as to have occasioned several acts of the legislature to inforce their use. At the same time, the proprietors and drivers of carriages appear to be convinced by experience, that a narrow-wheeled carriage is more easily and speedily drawn by the same number of horses than a broad-wheeled one of the same burthen.
>
> It is no wonder therefore that private interest, operating against public good, should have suggested expedients to elude the restrictions individuals were laid under; or that the letter of the law should be more attended to than the spirit of it.[1]

The real need was for better roads, rather than unenforceable regulations. The answer seemed to lie with the steadily increasing number of Turnpike Trusts, but the much-vaunted Turnpikes were often little better than the roads they replaced, particularly in the midlands and north where the new manufacturing interests were beginning to grow. Here is Arthur Young on the Wigan turnpike:

> I know not, in the whole range of language, terms sufficiently expressive to describe this infernal road. To look over a map, and perceive that it is a principal one, not only to some towns, but even whole counties, one would naturally conclude it to be at least decent; but let me most seriously caution all travellers

[1] J. Jacob, *Observations on the Structure and Draught of Wheel-Carriages* (1773).

who may accidentally purpose to travel this terrible country, to avoid it as they would the devil; for a thousand to one but they break their necks or their limbs by overthrows or breakings down. They will meet with rutts which I actually measured four feet deep, and floating with mud only from a wet summer; what therefore must it be after a winter?[1]

Even the better turnpikes were badly adapted for heavy commercial traffic: they were the routes of the post-chaise and the stage-coach. However, the reality of stage-coach travel was a very long way from the popular Christmas Card picture – all crisp, white snow, post horn and jolly punch-swilling landlord beside a glowing log fire at the end of the journey. Here is a touch of the reality:

I here learnt that the stage was to set out that evening for London, but that the inside was already full; some places were however still left on the outside. . . . I determined to take a place as far as Northampton on the outside.

But this ride from Leicester to Northampton I shall remember as long as I live. . . .

The getting up alone was at the risk of one's life; and when I was up, I was obliged to sit just at the corner of the coach, with nothing to hold by but a sort of little handle fastened on the side. I sat nearest the wheel, and the moment that we set off, I fancied that I saw certain death await me. All I could do, was to take still faster hold of the handle, and to be more and more careful to preserve my balance.

The machine now rolled along with prodigious rapidity over the stones through the town, and every moment we seemed to fly into the air; so that it was almost a miracle that we still stuck to the coach, and did not fall. We seemed to be thus on the wing, and to fly, as often as we passed through a village or went down a hill.

At last, the being continually in fear of my life became insupportable, and as we were going up a hill, and consequently

[1] Arthur Young, *A Six Months Tour Through the North of England*, 2nd edn 1770).

proceeding rather slower than usual, I crept from the top of the coach, and got snug into the basket. . . . As long as we went up hill, it was easy and pleasant. And, having had little or no sleep the night before, I was almost asleep among the trunks and the passages; but how was the case altered when we came to go down hill; then all the trunks and parcels began, as it were, to dance around me, and every thing in the basket seemed to be alive; and I every moment received from them such violent blows, that I thought my last hour was come. . . . I was obliged to suffer this torture nearly an hour, till we came to another hill, when quite shaken to pieces and sadly bruised, I again crept to the top of the coach, and took possession of my former seat. . . .

We at last reached Northampton, where I immediately went to bed, and slept almost till noon, resolving to proceed to London in some other stage coach, the following morn.[1]

He may have had a decidedly uncomfortable passage, but at least he did reach his destination. Brief news items like the following were commonplace in the newspapers of the time: 'On Sunday night between eleven and twelve o'clock, as the Salisbury stage was returning to London, it was overturned into a ditch near Basingstoke, owing to the coachman's being very much in liquor.'[2]

Land transport was slow, unreliable and expensive, and, in the case of bulky goods, such as coal and ore, could even be impossible. The alternatives were the coastal routes and river navigations. During the first part of the eighteenth century, a great deal of effort was put into improving river navigations, and the main rivers – Thames, Severn, Trent and Mersey – were busy with trade. But they still left the central areas untouched and, even at their best, suffered from severe disadvantages. In spite of improvements brought about by the construction of locks and weirs

[1] Charles P. Moritz, *Travels through Various Parts of England in 1782*, in William Mavor, *The British Tourists* (1809).
[2] *Felix Farley's Bristol Journal*, 16 August 1783.

and navigation cuts that by-passed the worst stretches, travel by river was not without its hazards:

> The barge *Emma*, laden with copper, and proceeding down the stream of the Thames, with all the expedition that could be used, was sunk on the 21st of February last on a shoal near Bell Weir Lock. . . . The same barge on *her return* being laden with grocery, &c. was *again sunk* on the 20th of March last, a little above Boulters Lock, having first been dragged over a shoal between Windsor and Maidenhead! ! ![1]

Traffic by means of these river navigations often involved trans-shipment, for there was very little cooperation between the bargemen of different rivers. Any attempt to encroach on a rival's territory could produce a fierce response, as William Darvall of Maidenhead discovered when he tried to take his Thames barges onto the Kennet. The local bargemen soon made their view of the matter very clear:

> Mr Darvall wee Bargemen of Redding thought to Aquaint you before 'tis too Late, Dam You, if y. work a bote any more at Newbery wee will Kill You if ever you come any more this way, wee was very near shooting you last time, we went with to pistolls and was not too Minnets too Late. The first time your Boat Lays at Redding Loaded, Dam You, wee will bore holes in her and sink her so Dont come to starve our fammeleys and our Masters . . . so take warning before 'tis too late for Dam you for ever if you come wee will doo it – from wee Bargemen.[2]

Unsatisfactory though they were, the river navigations were still a far more efficient and cheaper way of moving goods than any overland method. The rivers, though, were only connected to each other by the coastal traffic and this had its own difficulties. The most obvious of these was dependence on the weather – a sailing ship could be stuck in harbour for weeks waiting for a favourable wind and even when it left, there was no

[1] Western Union Canal, *Replies to the Arguments of the Thames Navigation Commissioners* (1820).
[2] Letter dated 10 July 1725 (British Transport Historical Archives).

guarantee of a speedy voyage to its destination. Sir Edward Parry, giving evidence before Parliament in favour of cutting the Caledonian canal, reported the case of two boats leaving Newcastle on the same day: the first reached Bombay before the second had completed its trip round the north of Scotland to Liverpool. There were other difficulties: whenever the country went to war, coastal vessels and their crews were always liable to be requisitioned by the Navy. The final, and probably the most irritating problem, was theft and pilfering, which affected both the ports and the river navigations. J. Phillips, a well-known canal propagandist, stated the case at its most extreme. Even so, his version reflects a view very widely held in the eighteenth century. He describes the practice of 'pilfering earthen wares, and other small goods, and stealing and adultering wines and spiritous liquors'. He goes on:

> The losses, disappointments, and discredit to the manufacturers, arising from this cause are so great, that they frequently choose to send their goods by land at three times the expence of water carriage, and sometimes even refuse to supply their orders at all, rather than run the risk of forfeiting their credit, and submitting to the deductions that are made on this account.[1]

So, by the 1760s there was an increasing demand for a reliable, efficient, cheap and speedy transport system. It also seemed obvious to a great many people that whoever produced such a system and ran it stood to make a great deal of money. So, among the ladies and gentlemen who watched the progress of the Duke of Bridgewater's canal, there were many who looked with more than disinterested curiosity. They had been told by some very authoritative people that the whole scheme was an expensive farce. A leading engineer of the day, on being shown the site for the proposed aqueduct at Barton that was to carry the canal over the River Irwell, was heard to remark that he had 'often heard of castles in the air, but never was before shown where any of them were to be erected'. But the 'castle in the air' was built and was

[1] J. Phillips, *A General History of Inland Navigation* (1792).

successful, and the interested spectators were not slow to draw the lesson from what they saw:

> The difference in favour of canal navigation was never more exemplified, nor appeared to more full and striking advantage than at Barton-bridge, in Lancashire, where one may see, at the same time, seven or eight stout fellows labouring like slaves to drag a boat slowly up the river Irwell, and one horse or mule, or sometimes two men at most, drawing five or six of the duke's barges, linked together, at a great rate upon the canal.[1]

Even more effective than the inspiring sight of the canal was the inspiring sight of reduced transport costs. Before the Bridgewater canal opened, coal sold in Manchester at 7d per hundredweight: the canal trade cut the price exactly in half. Manufacturers and traders, seeing such a dramatic drop in costs, became converts to the idea of canal transport, speculators with money in their pockets began to dream of fat cash profits, and the small group of men who had been engaged in building the Bridgewater canal found themselves suddenly in demand as the only experts in the new technology. Thanks largely to the initiative taken by the Duke of Bridgewater, the great age of canal construction was begun.

[1] *Ibid.*

2

THE FATHER OF CANALS

The canal of the great duke of Bridgewater, who may justly be called the parent and founder of all similar works in this kingdom, is a very striking instance of public utility . . . and no doubt the vast fortune which this noble adventurer thus sacrificed for the good of his country, at a time of life when others squander their patrimony in useless dissipation, will amply be repaid.

Rev. S. Shaw, *A Tour to the West of England in 1788* (1789).

By the middle of the eighteenth century, the need for improvements in transport was widely recognized. There was a limit to the amount that could be done in improving river navigation, so it was natural that the idea of canal construction should be widely aired. Various schemes were put forward, including an early plan for a waterway to connect the rivers Trent and Mersey, but for a variety of reasons none of them ever amounted to anything, and most never developed beyond the stage of being vague suggestions. The promoters of the different schemes were unsure of the practicability of their own plans, worried by the thought of the inevitable conflicts with landowners and rival transport interests, and unable to raise enough enthusiasm or, more importantly, capital to make a really strenuous effort to push their schemes through. Men were cautious about committing themselves to a new and totally untried enterprise. If canal development was ever to get under way, then someone had to take the initiative to start a scheme, and then had to keep on with it to a conclusion. The most likely candidates for the role of innovator were the new industrialists, the men who stood to gain most by such a development and who were comparatively free of the traditional conservatism of the older landed interests. Among the least likely candidates would be a nobleman, such as Francis Egerton, third Duke of Bridgewater.

Francis Egerton was born on 21 May 1736. He could claim a fair number of illustrious ancestors; his great-great-grandfather, Sir Thomas Egerton, had been Lord Keeper of the Seal at the court of Elizabeth I and became a highly successful Lord Chancellor under James I; his great-grandfather, John Egerton, had been the first Earl of Bridgewater, and had earned his place in the history books partly by his encouragement of the arts, and particularly by his association with the production of John Milton's *Comus*. The first Duke of Bridgewater was Francis Egerton's father, and when he died in 1745 the title passed to the eldest son. But the Egertons were sickly children: four of the brothers died and the title came to Francis when he was only twelve years old. Francis appeared to be no more robust than the rest of the family and there seemed to be every possibility that the line would end there, but he survived the bad health of his childhood.

The young Duke received the conventional education of his class, ending up with his being sent, at the age of seventeen, on the Grand Tour of Europe. He was fortunate in having as his travelling companion the scholar and antiquarian, Richard Wood, who saw to it that the young man learned something from his travels. They saw all the sights that the Grand Tour had to offer, and, just like the other tourists, the Duke came away from the sites of the ancient world weighed down with antique busts and statuettes, and paintings and sketches of the more romantic of the Roman and Greek ruins. His enthusiasm for antiquities failed to survive beyond the end of the Tour – the European treasures were packed into cases, delivered back to England and left unopened and neglected until the day he died. But it was not just antiquities that he went to see. He also visited the Grand Canal of Languedoc, now known as the Canal du Midi. This great waterway, 150 miles long, had been completed in 1681 and was an immense success. The Duke was impressed by what he saw of it, and this impression did stay with him beyond the end of the Tour. However, there is no evidence that he made any detailed inquiries about the practicalities and techniques of canal construction.

Back in England, the Duke naturally enough turned to the pleasures of the capital. He took his place in Society and then fell in love. The lady was the beautiful young Duchess of Hamilton. They were as unlikely a pair as could be imagined – the young Duke, very serious, rather straight-laced and puritanical, the Duchess, deeply involved in the intrigues of Society and already the subject of a certain amount of gossip. Inevitably, they quarrelled – not, surprisingly, over any of the rumours concerning the lady herself but about her sister, Lady Coventry, who had been involved in what was for those times a not-particularly scandalous affair. However, the mild scandal was more than the Duke was prepared to tolerate and, rather self-righteously and unreasonably, he demanded that the Duchess of Hamilton should publicly disown her sister. The lady refused and that was that. The Duke left London in 1758 and the whole direction of his life turned dramatically. He 'is said never to have spoken to another woman in the language of gallantry. A Roman Catholic might have built a monastery, tenanted a cell, and died a saint. The Duke, at the age of twenty-two, betook himself to his Lancashire estates, made Brindley his confessor, and died a benefactor to commerce, manufacture, and mankind.'[1]

In Lancashire, the Duke busied himself in discussions with his agent, John Gilbert, and the two men occupied their time in planning ways in which the coal trade from the Worsley estates could be improved. This in itself was not particularly incongruous – many aristocrats had become enthused with the idea of finding mineral wealth under their lands. It was not unheard of for the grounds of a country estate to be alive with workmen tearing up the lawns and flower beds in the search for coal. It

[1] Earl of Ellesmere, *Essays* (1858).

was, however, unusual to find such a young man occupying himself with the problems of trade.

The coal trade from the Duke's mines at Worsley did offer a problem. Recently, the navigation of the near-by River Sankey had been transformed by the construction of a cut parallel to its route, the Sankey Brook, and this improved route to other collieries had resulted in part of the Duke's old market being taken away from him. He badly needed a new outlet for his coal, and for this he needed cheap transport. The most obvious outlet was the rapidly-growing town of Manchester. Population estimates for the period are not particularly reliable but they at least indicate the rate of development. In 1717 the population was estimated at 8000. In 1757 'the number most to be confided in, for Manchester and Salford' is 19,839 and by 1773 a survey 'executed with accuracy' gave a total of 42,927.[1]

The Duke's first thought was to construct a short cut from Worsley to Barton where he could get access to the River Irwell and thence to Manchester, but the proprietors of the Mersey and Irwell Navigation Company refused to negotiate reasonable terms. Forced by the intransigence of the old Navigation Company, the Duke and Gilbert took the vital decision – to construct an independent navigation which would run direct from the colliery to Salford, just outside Manchester.

This was a bold decision for the Duke to take. No one in Britain had ever attempted a cross-country canal of this type, so there were no engineers with the necessary experience to direct the proceedings. There were no precedents to follow, no prototypes to copy. True, the Duke knew from his continental travels that such a canal could be constructed, but he certainly did not know how to set about it. Ignorance of the techniques was only one of the problems he had to face. Such a scheme would undoubtedly be very expensive, and while he was in no way a poor man, he would be committing himself to, literally, incalculable expenditure. He also knew that he would have to face opposition from a number of well-organized and powerful trading interests,

[1] J. Aiken, *A Description of the Country from thirty to forty miles around Manchester* (1795).

notably the Mersey and Irwell Company. This was precisely the combination of difficulties that had deterred other would-be canal constructors. It says a great deal for the single-minded character of the Duke that he was prepared to face these difficulties, not as one of a group or as part of a Company, but on his own, depending entirely on his own resources. And he was still only twenty-two years old.

The first struggle that the Duke faced was to get an Act through Parliament to enable the work to begin.[1] However, the Duke was successful, and obtained his Act in March 1759.

In the summer of that year, the Duke was introduced by Gilbert to James Brindley, a millwright, who was gaining a reputation as an ingenious mechanic and engineer. Brindley was employed as the engineer for the canal, and work could begin. Once it was under way, the Duke had time for nothing else: it dominated his whole life. The original plan was dropped for a more ambitious one that would take the canal right into Manchester, and this plan, in turn, was enlarged by the idea of extending the canal so as to join with the Mersey at Runcorn. Both proposed changes involved yet more arguments and fights, until the modifications were approved by two further Acts of Parliament in 1760 and 1762.

The greatest problem facing the Duke was money. As with many a canal scheme that was to follow his own, he found that costs rose continuously above original estimates. But where later canal builders could appeal to the general public and to subscribers for fresh funds, the Duke had only his own resources to draw on. He began to mortgage his estates. He no longer had either interest in or use of his London house, so that was disposed of. He raised what he could on his country estates, but still he was short of money. He borrowed money from his relatives, including the canal enthusiast Lord Gower; he borrowed from Manchester manufacturers who would stand to gain by the construction of the canal; and he borrowed from the Child and Co.

[1] The form which this and subsequent Parliamentary battles took will be dealt with later.

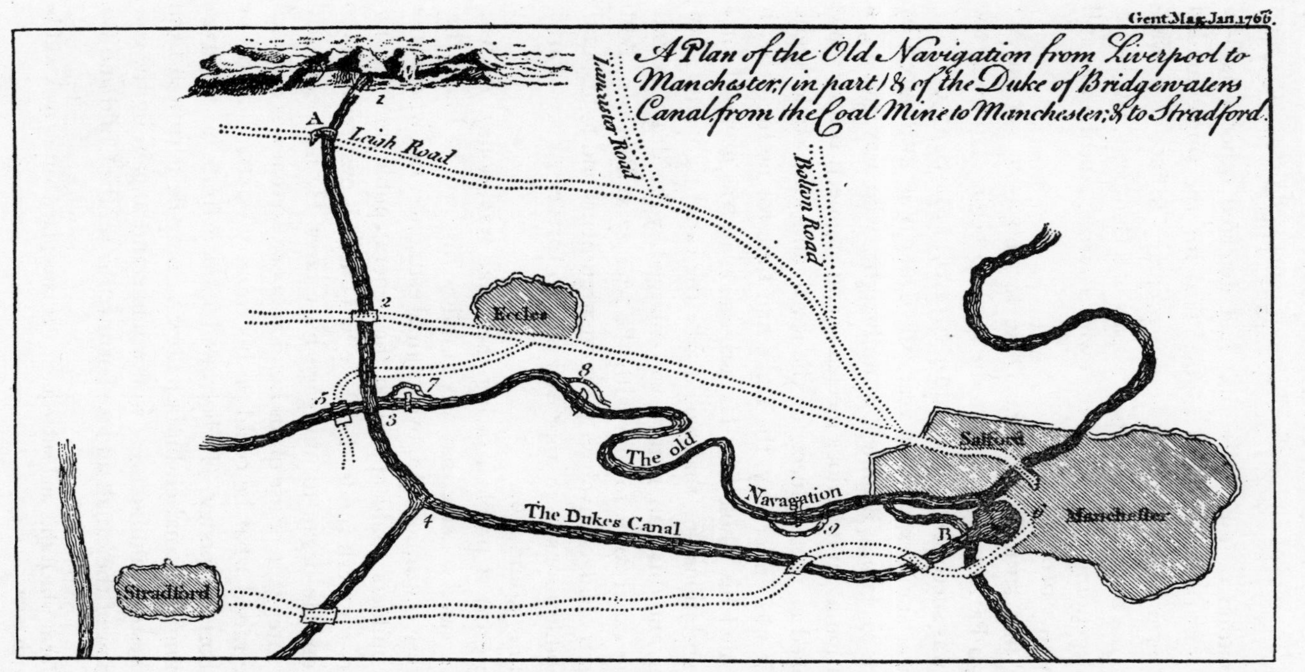

1 The Bridgewater Canal from Worsley to Manchester and part of the Irwell Navigation.

bank. In spite of all his efforts, funds remained desperately short. It became increasingly difficult to get credit anywhere. At one time his bill for £500 could not be cashed in the Liverpool area, and his long-suffering agent, Gilbert, spent a good deal of his time riding round the local farms, borrowing what he could where he could. The Duke was once actually pursued by the local parson who was trying to collect on a debt, and had to suffer the ignominy of being finally collared in a hay loft. But it was not just the lack of funds that must have weighed heavily on the young Duke: there was the general lack of confidence in the whole scheme which he was constantly having to face. For reassurance, he had only the small group who worked with him on the project. Many years later, Sir John Rennie told of a meeting between his brother and Mr Bradshaw, the Manager of the Bridgewater Canal:

> Pointing to a little whitewashed house, near the Moss, about half a mile distant, he said to my brother: 'Do you see that house? Many a time did the late Duke of Bridgewater, Brindley, and myself spend our evenings there during the construction of the canal, after the day's labours were over; and one evening in particular we had a very doleful meeting. The Duke had spent all his money, had exhausted his credit, and did not know where to get more, and the canal was not finished. We were all three in a very melancholy mood, smoking our pipes and drinking ale, for we had not the means to do more, and were very silent. At last the Duke said: "Well, Mr Brindley, what is to be done now?" Brindley said: "Well, Duke, I don't know; but of this I feel as confident as ever: if we could only finish the canal, it would pay very well, and soon bring back all your Grace's money." '[1]

Brindley's confidence was well placed. Somehow the money was scraped together, the canal was completed, and the completion was a triumphant vindication of the young man's dream. In the famous portrait done of him at the time we see him posed against a background of the canal, pointing with justifiable pride

[1] Sir John Rennie F.R.S., *Autobiography* (1875).

to its busy traffic. The time and effort he had put into the work finally began to show their reward; the canal was a financial success and the Duke's worries began to recede. By 1769 he had repaid the whole of the £25,000 that he had borrowed from Child's bank. He was still young and handsome, and now he was wealthy again. Had he wanted to do so, he could have returned to the fashionable world, but by now his work with canals had become an obsession. The rest of his life was to be spent mainly in the north-west, working on improvements to his canal, lending his support to other canal promoters and concerning himself with his colliery and his lands.

The Bridgewater Canal itself was a broad, level waterway continuing without locks from Worsley into Manchester. At Worsley, it disappeared underground into the mine workings, the drainage from which provided the main source of water for the canal. Over the years, the underground workings became more and more extensive, an elaborate network of subterranean waterways that stretched even farther than the canal itself. At the coal face, the coal could be loaded almost directly into the boats:

> The coals are brought to this passage or canal in little low wagons, that hold nearly a ton each; and as the work is on the descent, are easily pushed or pulled along, by a man on a railed way, to a stage over the canal, and then shot into one of the boats, each of which holds seven or eight tons. They then, by means of the rails, are drawn out by one man to the basin at the mouth, ... then five or six of them are linked together, and drawn along the canal by a single horse, or two mules on the banks or towing-paths.[1]

The extension to Runcorn was taken down to the level of the Mersey by an impressive flight of locks that excited almost as much admiration as the famous aqueduct at Barton had done. The potter Josiah Wedgwood came to see them and recorded his impressions in a letter of 21 June 1773:

[1] J. Phillips, *op. cit.*

> We set our faces for Runcorn Gap . . . to behold the wonderfull works of his Grace of Bridgewater, & truly wonderfull they are indeed. You know I have seen a good deal of these matters before, but notwithstanding that, I was quite astonished at the vastness of the plan, & the greatness of stile in the execution. The Walls of the Locks are truly admirable, both for strength, & beauty of workmanship. The front Lock next to the Sea (for such it seems when the Tide is in) in particular, whose walls are compos'd of vast stones from 1 to 12 Tons weight, & yet by the excellent Machinery made use of, some of which is still left standing, they had as perfect command of these huge Masses of Rock, as a common bricklayer has of the brick in his hand . . . the whole seems to be the work of Titans, rather than a production of our Pigmy race of beings.[1]

The Duke, Gilbert and the engineer Brindley worked on many other improvements. At the Manchester end of the canal, the terminus was built at the foot of Castle Hill, which meant that the poor who came to buy their coals were faced by a long and tiring trudge back up the hill with their loads. To overcome this, the canal was extended into the hill and a vertical shaft was sunk down into it from the hilltop. Above the shaft there was a hoist, powered by a water wheel. The coal, which had been loaded into the iron boxes at the coal face, could now be lifted in the same boxes up to the top of Castle Hill.

As well as concerning himself with his canal, the Duke also earned a reputation as a good manager of his estate, receiving the accolade of 'good husbandman' from no lesser authority than the famous agriculturist Arthur Young. But always it was for his canals and his canal schemes that he reserved his real enthusiasm, to such an extent that he came to be regarded as something of an eccentric by his contemporaries:

> In person he was large and unwieldly, and seemed careless of his dress, which was uniformly a suit of brown, something of the cut of Dr Johnson's. Mr —— of ——— passed some days with us, and during his stay the duke was every evening

[1] Josiah Wedgwood, *Letters 1772–1780* (1903).

planted with him on a distant sofa in earnest conversation about canals, to the amusements of some of the party.[1]

The Duke undoubtedly had little interest in the currently fashionable preoccupations or in the exchange of tit-bits of aristocratic scandal. But, however enthusiastic he might be, he never allowed his enthusiasm to deter him from considering new ways of raising money from his canal. Like many an aristocratic property owner of a later age, he had a keen eye for a profit. The canal was as popular as a tourist attraction as it was as a commercial enterprise, and he soon found a way of making money out of its attractions:

> The Duke of Bridgewater has just built two packet-boats, which are every day towed from Manchester to Warrington; one carries six score passengers, the other eighty: Each boat has a coffee-room at the head, from whence wines, &c. are sold out by the Captain's wife. Next to this is the first cabbin, which is 2s. 6d. the second cabbin is 1s. 6d. and the third cabbin 1s. for the passage or voyage upon the canal.[2]

A happy combination of useful passenger traffic and pleasure trip!

As the canal prospered and the duke grew older, he was able to relax, but still spent most of his time wandering his estates, rarely travelling far from the house he had built overlooking his canal. His manner was rough, but he seems to have left behind a reputation as a good and honest employer, and he showed more compassion than many of his contemporaries: 'During the winter of 1774 the Duke of Bridgewater ordered coals to be sold to the poor of Liverpool in pennyworths, at the same rate as by cart loads. Twenty-four pounds of coal was sold for a penny.'[3] He also found his way into popular ballads as a supporter of the hand-loom weavers against the factory owners' attempts to introduce power looms:

1 Earl of Ellesmere, *op. cit.*
2 Letter from Warrington, 1 September. *Annual Register*, 1774.
3 Thomas Baines, *History of the Commerce and Town of Liverpool* (1852).

1 Francis Egerton, Third Duke of Bridgewater, points with pride at th
Barton aqueduct and the labouring teams hauling loads on the Irwe

2 'The Wedgwood Family' by George Stubbs. Josiah is sitting on the bench.

3 Wedgwood's Etruria works, with the Trent and Mersey Canal in the foreground.

For coal to work his factory
He sent unto the Duke, sir;
He thought that all the town
Should be stifled with the smoke, sir;
But the Duke sent him an answer,
Which came so speedily,
That the poor should have the coal,
If the Devil took th' machinery.[1]

The Duke was not, however, a man blessed with the sweetest of tempers. On one occasion, he returned from one of his comparatively rare trips to London to find that the gardeners, in an attempt to bring a little more elegance to the grounds of his Cheshire home, had planted new flower beds outside the windows. He was not pleased. When he looked out of his windows what he wanted to see were canal, docks and wharves – not flowers. He stomped out of the house, cane in hand, and set about the systematic demolition of the offending plants.

The Duke of Bridgewater died in 1803, having lived long enough to see the canal system that he had begun spread to all parts of the kingdom. He did not, however, live long enough to see another of his prophecies fulfilled: Lord Kenyon had congratulated him on the success of the canal – 'Yes,' replied the Duke, 'we shall do well enough if we can keep clear of those damned tramroads.'

[1] Part of a ballad, 'Grimshaw's Factory Fire', written by a hand-loom weaver called Lucas, and quoted in John Harland, *Ballads and Songs of Lancashire* (1856).

3

THE PROMOTERS

'Twas just at the time when in sorrowful strain,
Old England was grievously groaning,
Her natives in sadness, to add to the scene,
the loss of their trade were bemoaning:
To give some redress, in this age of distress,
Some worthies (tho' few in the nation)
As a scheme that might tend, to some favourable end,
Were resolved to promote navigation.

In Lancashire view what a laudable plan,
And brought into fine execution
By Bridgewater's duke; let us copy the man,
And stand to a good resolution:
If the waters of Trent with the Mersey have vent,
What mortal can have an objection!
So they do not proceed, to cut into the Tweed,
With the Scots to have greater connection.

A free intercourse with our principal ports,
For trade must be certainly better;
When traffic's extended, and goods easy vended,
In consequence things will be cheaper:
Our commerce must thrive, and the arts will revive,
Which are now in a sad situation;
If we follow this notion, from ocean to ocean,
To have a compleat navigation.

To the land what advantages soon must proceed,
When once we have opened our sluices?
Our cattle, and even the land where they feed,
Will be turn'd into far better uses:
'Tis this will enable our merchants abroad
To vie with each neighbouring nation:
Who now, as they tell us, in fact undersell us
For want of this free navigation.

'Inland Navigation', a song to the tune of the Marquis of Granby, *Gentleman's Magazine*, March 1766.

This poem is quoted not because it is a very delightful piece of jingle writing, which it is, but because it reflects so accurately the mood that started to spread through the country once the success of the Bridgewater Canal was established. The light-hearted optimism, the appeal to patriotism, even the rather snide references to the Scots – a reminder that Jacobitism was neither forgotten nor forgiven – were all typical of the period. But the keynote is the feeling that the expansion of canals brought a promise of an almost unlimited expansion of trade, and it was this feeling that brought men together to blow the dust off old canal plans or to dream of new ones. The 'worthies' might, in the 1760s, still be 'few in the nation', but they were beginning to make their presence felt as they began the lengthy business of promoting the new schemes.

The job of the canal promoters was first to raise local interest and enthusiasm, to get agreement on the general plan, and then to organize enough money and support to give the plan a reasonable chance of getting through Parliament. The men who undertook this task as prime-movers fell into two broad categories. There were those who, like the Duke of Bridgewater, came from the old landed families, and there was a second, and larger, group made up of the new class of industrialists and traders. In most canal promotions, both groups were represented.

It would be tedious, and not particularly profitable, to look in detail at a whole string of canal promoters and their schemes. They tended to follow a definite pattern, which was established with the earliest canal promotions. Instead, this chapter will be mostly concerned with the promotion of one particular scheme, that for the Trent and Mersey – or Grand Trunk – canal. The promotion was the first to follow after the Bridgewater, and is therefore both interesting and important, and in its leading proponent, Josiah Wedgwood, had one of the most fascinating of the eighteenth-century industrialists.

The history of the Trent and Mersey began in 1755, when a scheme was put forward for a canal to join the Trent and the Severn. This, however, was quickly replaced by a second plan, and a survey was made for a canal that would link together Chester, Stafford, Derby and Nottingham. No progress was made and the matter rested until Earl Gower became interested. He already had connections with canal construction through his brother-in-law, the Duke of Bridgewater, and he commissioned a fresh survey of the line and employed James Brindley for the job. The survey was made between 1759 and early 1760, but there was no public support, and everything was dropped again until the opening of the Bridgewater canal produced a strong revival of interest.

The main impetus behind the revived plan came from the potters of Staffordshire, with Josiah Wedgwood as their leader, and the traders and merchants of Liverpool. Earl Gower was, naturally enough, involved, but the active day-to-day business of the canal promotion fell largely on Wedgwood.

The story of Wedgwood's life could almost be taken as a model to illustrate the development of an eighteenth-century industrialist: not exactly rags to riches, but certainly from modest means to very considerable wealth and influence.[1]

He was born in 1730, the thirteenth and youngest son of Thomas and Mary Wedgwood, who in turn were part of a large family spread out across Staffordshire and including many master potters and journeymen. When he was nine years old, his father died and Josiah was taken out of school and apprenticed to

[1] The main source of information on Wedgwood's life is the two-volume biography by Eliza Meteyard, published in 1865, and the Wedgwood letters, published in two volumes, 1903 and 1906.

his brother Thomas. He began to train as a thrower, learning to produce the crude earthenware pots that were still the staple trade of the potteries, but at the age of twelve he was attacked by small-pox, which left him with a badly weakened right knee, and the strain of working the treadle for the potter's wheel proved too great. At fifteen, he gave up throwing and began the learn other facets of the potter's craft. He started to experiment with ways of improving the quality of the pot and the glaze, but his brother was unimpressed. Thomas had a comfortable living, was happy with the old ways and regarded experiments as a frivolous waste of useful time. So, when Josiah's apprenticeship was over, Thomas declined to take him into the partnership. Josiah went instead to another potter, in Stoke, who was delighted to make use of the young man's suggested improvements in technique. He was also delighted by the increased profits that followed the improvements but, unfortunately, as none of the increased profits reached the innovator himself, that particular partnership lasted only a year. Josiah moved on around the potteries, gaining in experience and saving what he could, until he had enough capital to set up in business for himself.

Two distant Wedgwood cousins, John and Thomas, leased their Ivy House works to Josiah. For the cottage, workshop and two kilns, Josiah paid £10 a year in rent and, with another cousin Thomas as journeyman, he began business in the early part of 1759. From the very beginning, his aim was to improve the standard of English pottery to bring it to a state where it could compete with the work coming across from such continental centres as Delft. At first he had to content himself with making small pieces, or replacements of single pieces of high quality, when a plate or cup that belonged to a continental set was broken. In these early days, he had to do almost everything for himself, but still found time to pursue his main purpose. He wanted the aristocracy to come to Staffordshire for their pottery, so he concentrated on improving the very expensive cream ware, confident that once he had aristocratic backing, he would be able to increase production and bring down prices low enough to put Wedgwood china within the reach of the middle classes as well.

His success was assured when he at last received an order from the Queen – the rest followed exactly as he expected.

Wedgwood's search for better quality and increased production forced him to reorganize his workshops: the Wedgwood works began more and more to resemble a modern factory, with increasing specialization among the work force. But his attempts at improvement were always being hampered by the transport difficulties that plagued many of the industrialists. For his new cream ware he needed special clays from the West country – and with land carriage at the exorbitant rate of 9s per ton for a ten-mile journey and water-carriage cheaper at 3s 4d a ton, but less reliable, transport was a major headache, and its problems occupied too much of Wedgwood's time. The business did increase rapidly in the early 1760s, but the increased business also meant an increase in these transport difficulties. Wedgwood needed greater supplies of raw material, and he was also producing more of the finished product, and it was the delicate china that suffered most from being humped around overland. If England's roads were bad, those of the potteries were probably worst of all – for years local potters had made up shortages of clay by digging it up out of the highway outside their workshops. Wedgwood found himself constantly having to ride between his works and the port at Liverpool to oversee the transport arrangements. He was a man of immense vigour, as he needed to be to survive the 'health courses' that he set for himself – 'riding on horseback for 10 to 20 miles a day, & by way of food & Physick, I take whey, & yolks of Eggs in abundance, with a mixture of Rhubarb & soap, & I find the regimen to agree with me very well'.[1] But in spite of his 'agreeable' regimen, the tiring journey led to a recurrence of the old trouble with his knee and on one of his Liverpool visits, early in 1762, he was forced to retire to his bed. His doctor, seeing how irritated the enforced rest was making Wedgwood and realizing that unless he found some diversion for him his patient would not stay in bed for long, introduced Wedgwood to Thomas Bentley, a Liverpool merchant.

Bentley was an educated man, and, rather more important, an

[1] Letter dated 6 July 1767.

intelligent one – his house in Liverpool was a regular meeting place for the intellectuals of the district. The doctor had chosen well and the two men became friends. Their conversation naturally included discussions on problems of trade and transport, and when Wedgwood returned to the potteries, they continued to correspond on the subject. The revival of interest in the Trent and Mersey canal scheme found them both already converts to the idea. It seemed natural for Wedgwood to become the main promoter of the scheme in the Staffordshire area, while Bentley would handle the Liverpool end. As time went on, Bentley's role began to become more and more subservient to Wedgwood's.

Wedgwood's task was a dual one – to drum up enthusiasm for the scheme so that when eventually the time came to present the plans to Parliament they would be assured of a solid body of support, and to attempt to deal with objectors and opponents. Even before any detailed plans were drawn up, Bentley was hard at work as a pamphleteer, expressing their case. Wedgwood and friends acted as critics – and one friend, Dr Erasmus Darwin, was more critical than most. Some of his criticism was general and sensible: 'it is not wrote enough to the landowners, & they, if any body of men does, will ruin our scheme,' he wrote in September 1765, but some criticism was just pernickety. Wedgwood found himself acting as unwilling mediator between the increasingly critical Darwin and the increasingly irritated Bentley. Eventually, Wedgwood lost his patience too, as yet another delay for correction was proposed: 'the condemnation or postponing of the whole I can by no means agree to nor persuade myself that there is any necessity for it. Must the Uniting of Seas & distant countries depend upon the choice of a phrase or monosyllable? Away with such hypercriticism, & let the press go on, a Pamphlet we must have, or our design will be defeated, so make the best of the present, & correct, refine, & sublimate, if you please, in the next edition.'[1] Wedgwood, one feels, had the making of a successful publisher! His irritation on this occasion was no more surprising than Bentley's, for Wedgwood too was furiously busy: 'I have not time to think or write about anything, but *the*

[1] Letter to Bentley, 15 October 1765.

immediate business of the day, Public business I mean, for as to my private concerns I have almost forgot them, I scarcely know without a good deal of recollection whether I am a Landed Gentleman, an Engineer or a Potter, for indeed I am all three & many other characters by turns, pray heaven I may settle to something in earnest at last.'[1]

Wedgwood was finding that canal business took up more and more of his time: A letter to Erasmus Darwin[2] gives an idea of just how busy he was:

> We met Mr Gilbert in our way to Trentham & stp'd here long enough to shew him our plan and explain the scheme a little to him – He immediately ask'd if it could not join the Duke's Canal. . . . Councillor Gilbert was at Trentham & was highly pleas'd with the Plan and scheme. . . . By this day's post I shall write to the Mayor of Liverpool to inform him what is in agitation in this part of the world & to beg leave to lay a plan &c. before him. And I am going to wait upon Mr Willbraham of Rhode with a plan &c. to day.

He was ceaselessly travelling the country to raise support, and to argue for his scheme and against that of various rivals. The old Navigation Company of the River Weaver represented the main challenge, and they put forward various schemes of their own – one was for a new canal that would by-pass the potteries completely, and another was for a waterway that would surround the existing Bridgewater canal, and so make expansion eastwards virtually impossible. But Wedgwood found he had almost as many problems with friends and supporters as he did with rivals. Thomas Gilbert, M.P., the brother of John Gilbert, the Duke of Bridgewater's agent, put up a scheme that 'astonish'd, confounded and vexed' Wedgwood. Gilbert's proposal was for the management of the canal to be taken out of the hands of the proprietors and placed with a body of Commissioners who would, among other duties, regulate the charges and control the property of the Company. The plan was taken to Earl Gower,

[1] Letter to Bentley, 2 January 1765.
[2] 15 April 1765.

and Wedgwood was summoned to a meeting at Lichfield in January 1765, where he attacked the Gilbert proposals. 'In 10 minutes he found his *baseless Fabrick* tumbling to the ground' and Wedgwood went off 'huzzaing and singing I o' after a conquest.'[1]

Wedgwood discovered that the diplomacy involved in canvassing for support occasionally demanded a little deceit. Bentley wrote to him about the Mayor of Liverpool: 'I have my Doubts whether he will enter heartily into the Business – and these Doubts arise from the *Facts* you collected at Northwich – about Coals which at present should be concealed as much as possible – Mr Tarleton's [the Mayor] Aunt Clayton is deeply interested in the Coal Works on the Sankey Navigation.' But the problems of Aunt Clayton's stake in a rival coal company seem to have been overcome, for in May 1765 Wedgwood was able to write to Darwin:

> On Friday last I set out for Liverpool & took a friend along with me, we arrived there the next day, thoro'ly dusted, & almost melted down to oyl . . . dined with his Worship the day following along with several eminent Merchants. . . .
>
> Mr Mayor introduced the subject after dinner very genteelly by giving for a toast – 'Success to a scheme for uniteing the Trent & Weaver by a Navigable canal'. The toast went round the board with glee, after which our plans were introduced by his Worship & the whole design explain'd to them in the best manner we were capable of doing it, & they seem'd so thoroughly convinc'd of its importance . . . that I am very certain we may depend on every assistance we can ask for that quarter.[2]

But jealousies and rivalries arose. The Lancashire subscribers decided to survey their end of the canal for themselves and run it as an independent unit paid for out of their own subscriptions, so that they could insure against any possible damage to Liverpool interests. Wedgwood lost no time in putting them very firmly in their place:

[1] Letter dated 2 January 1765.
[2] Letter to Darwin, May 1765.

You have strange heterodox notions amongst you at Liverpool about your Port being ruin'd, your not being principals, & I don't know what stuff. Pray who are *principals by the rules of common sense*, in a design of this sort, but those who will receive the *principal* advantages from it. . . . Pray now seriously what will your Lancashire Canal be without the Derbyshire, Staffordshire, & Cheshire additions – it dwindles so that I can scarce see its *importance* without the assistance of a *mental microscope*.[1]

The journeys that Wedgwood undertook to raise support were no light matter for him, for he was constantly suffering pain. Although at the time he was probably unaware of just how serious the trouble was, by the end of the decade his leg had to be amputated. The operation was, of course, performed without anaesthetics, but he sat in his chair watching the surgeons at work, and never uttered a sound. Even with that kind of stoic bravery, however, Wedgwood must have welcomed whatever help he could get. One of his supporters, who gave time to promoting the canal, was a local landowner, Sir William Meredith, and in return Wedgwood offered a little friendly advice on how to win an election that Sir William was fighting:

RECEIPT Bullocks roasted whole – Quantum sufficit.
6 small Cannon to be fired at every vote gained from the Enemy.
A Fighting Captain to be made use of occasionally with the wavering and timerous.
Get Mr Scaigs (a person well known in Tamworth) to make *quere faces*.
A Poet is absolutely necessary & may be heard of at Birmm.

As well as raising support for the canal, he had to assist in fixing its actual line and this involved many consultations with the engineer, Brindley, who had conducted the original survey, as well as going to see the various landowners along the route.

[1] Letter dated 14 Janary 1766.

Brindley, as many a canal proprietor was to discover, was not the easiest man to deal with, but he was well known in the potteries. He had established a flint mill in the Burslem area, and he and Wedgwood became friends: 'Mr Brindley & his lady call'd here in their way home, lay with us & are just left this morning. We are to spend tomorrow with them at Newchapel, & as I always edify full as much in that mans company as at Church I promise myself to be much wiser the day following.'[1] It was as well for Wedgwood that he did establish this rapport, for he had quite enough to do without becoming embroiled in quarrels with the engineer and surveyor.

Eventually, plans were finalized. The route selected was to join the Trent to the Mersey by means of a junction with the proposed extension of the Duke of Bridgewater's canal at Runcorn. The Trent and Mersey supporters were thus placing themselves in direct competition with the Weaver Navigation (see map, p. 179) who tried, belatedly and unsuccessfully, to reach some sort of compromise that would protect their interests. But the Trent and Mersey supporters were sure of their ground: their canvassing for support had been successful, they had no real fear about raising the necessary funds, and they were at last ready to call a public meeting, which was held at Wolseley Bridge:

> Earl Gower, Lord-Lieutenant of the County of Stafford; Lord Grey and Mr Bagott, Members for that county; Mr Ashton Curzon, Member for Clitheroe; Mr Anson, Member for the city of Lichfield; Mr Gilbert, Member for Newcastle-under-line, and many other of the principal gentlemen and landowners of that and the neighbouring counties, as well as several merchants and tradesmen from Liverpool, Birmingham, and other great trading towns, were present: Lord Gower opened the meeting . . . ever since he had heard of the scheme, it had been his determination to support it with all his interest, both provincial and political; for he was satisfied that the landed

[1] Letter to Bentley, 22 February 1768.

and trading interests were so far from being incompatible, that they were the mutual support of each other.[1]

The meeting was a triumph for the canal's supporters, and it was agreed to petition Parliament immediately to ask for a bill. Wedgwood was delighted, and the other potters of the district were no less enthusiastic:

> In the North of Staffordshire, the Potters are so sensible of the prodigious benefits that must accrue to their trade by this intended Canal, that on Tuesday Night many of them assembled together at Burslem, over a large Bonfire, and drank the Healths of Lord Gower . . . and other Well-wishers to the Navigation, with the loudest Acclamations of Joy.[2]

The first stage of the canal promotion was over. Wedgwood continued to take an active interest in the canal's progress, though he was never again involved in quite such hectic activity. There was one more period of frenzied work when the Parliamentary battle was fought (see pp. 39–40, 59–60), and after that Wedgwood's part in the proceedings was acknowledged by his being appointed Treasurer to the new Company. He was both pleased and proud. He wrote to his brother, John:

> We had yesterday a very Noble, Numerous & amicable General assembly of the Comm^{rs} & Proprietors of the *Navigation from the Trent to the Mersey* . . . which was conducted through the whole of it with the utmost order, Harmony & as far as appear'd to the entire satisfaction of all parties . . . you will see the honour done me, which was quite unexpected & voluntary, without the least previous sollicitation on my part, and without one dissenting voice.[3]

But the honour he received never made Wedgwood pompous – he retained a sense of humour. At the end of the letter he gave his brother a note on the officers appointed:

[1] *The Gazetteer and New Daily Advertiser*, 9 January 1766.
[2] *The Public Advertiser*, 9 January 1766.
[3] Letter to John Wedgwood, 4 June 1766.

James Brindley Surveyor General £200 per ann.
Hugh Henshall Clerk of the Works £150 per ann. for self & clerk
T. Sparrow Clerk to the Proprietors £100 per ann.
Jos. Wedgwood Treasurer at £000 per ann. out of which he bear his own expenses.

The appointment as Treasurer was a mark of how far Wedgwood's importance had grown and how well his business had prospered, for where a mere seven years ago he had just managed to raise enough cash to pay a rent of £10 per annum, he was now able to put up a security of £10,000.

In 1769 he opened his famous Etruria works on the bank of the canal and had a suitably grand home, Etruria Hall, built near by. It was from these works that the famous Frecian and Etruscan style vases and jasper ware were produced to be sent to all parts of the world. Wedgwood became an avid collector of books and prints: his experiments with pottery led him into more experiments in the field of chemistry, and he became so expert that the famous Lavoisier wrote to ask him to contribute to the *Annales de Chimie* and in 1783 he was elected a Fellow of the Royal Society. He also became involved in middle-class radical politics, as a keen advocate of Parliamentary reform and a supporter of Wilberforce and the campaign for the abolition of the slave trade. But, essentially, he remained a potter, proud of his own craftsmanship; at the opening of the Etruria Works, he sat down at the potter's wheel, and while his friend Bentley, who had been taken into the partnership, turned the wheel, he threw six pots to commemorate the event. It was said of Wedgwood that he never lost his eye for a good pot or a bad one – stomping round the works, he would raise the stick that he always used for support and bring it crashing down on any piece that failed to meet his exacting standards. But the 'simple potter' also became a very rich man, and when he died in 1795, he left a fortune of half a million pounds behind him.

Wedgwood and Earl Gower were, each in their way, typical of the type of men whose support was necessary to push through a successful canal promotion. Wedgwood was full of enthusiasm, immensely energetic, and, as a manufacturer himself, well able to convince other manufacturers and traders of the advantages of the waterway. He acted as an advocate not just for one particular canal, but for the whole idea of inland navigation and the expansion of trade. Earl Gower provided the one essential attribute that Wedgwood lacked – political power and influence among the landed interests. The long campaign that was waged to launch the Trent and Mersey was repeated many times over during the next half-century, but the lines of the development were laid down with the early schemes.

Not all canal promotions followed the same sequence. A few canals followed the Bridgewater pattern, and were built by individuals. Others were built by manufacturers to serve some particular local need. These would often be short canals, joining the factory to some other transport system. The iron makers of Shropshire built such canals to join their foundries to their main transport route, the River Severn, served by the docks at Coalport. The great Quaker families – the Darbys and the Reynolds – were well able to promote and finance canals from their own resources and to carry the whole through to a satisfactory completion. Lastly, as with the Caledonian canal, the promoter could be the government itself. But all these are the exceptions; the main job of canal promotion was carried out by the colliery owners, the textile manufacturers such as Arkwright and Strutt, the traders and the merchants, and by the few aristocrats, such as the Duke of Bridgewater and Earl Gower, who took up the cause of the canals.

4

THE OPPOSITION AND THE PAMPHLET WARS

When two opponents have said all that is true, they generally say something more; rancour holds the place of argument.

W. Hutton, *An History of Birmingham* (4th edition), 1809.

The opposition to canals came from many quarters. Probably the most important, and certainly the best organized, was that from the river navigations whose interests were threatened by the new developments. As canals spread and multiplied, the old navigations were in turn replaced and older canals became threatened by new schemes. The other most powerful force was the landowner, who might oppose a projected canal either from political conviction or because he felt it interfered with his private interests. Among this group were the mill owners, who were particularly concerned about any threat to their water supply. Finally there was an assortment of special and localized interests, men with personal prejudices to air, as well as a fair sprinkling of eccentrics.

The arguments between the river navigations and the canal promoters were very basic. The canals concerned were the first to be promoted, so that the battle was something more than just a disagreement between competing interests over specific routes – on the canal promoters' side it also had to be an argument for the whole idea of canal navigation.

The Duke of Bridgewater was the first to be involved in this particular battle. When he first began, his plans aroused no great interest among the river navigations. They refused to offer him any special terms, and then left him alone to get on with his plan. But when it became clear that he was planning not only a route for his own coal, but a rival transport system for general use, and when he decided to take his canal into the Mersey, that was a very different matter – the fight was on.[1] The Mersey and Irwell Navigation made a rather belated offer to reduce their exorbitant tonnage rates, but once their offer had been rejected, the argument was continued on both sides with very little regard for the niceties.

In many ways, it was an unequal struggle: during their monopolist days, the Mersey and Irwell proprietors had gained a bad reputation – high fees, delays, lack of repairs and maintenance. When they now tried to plead injured innocence, their pleas were a little unconvincing. The supporters of the Bridgewater canal were only too happy to remind the public of the past sins of the Mersey and Irwell, who had got up:

> an ill-timed petition, intended to be offered to you to sign, to give it (if not the reality) at least the appearance of popular sanction, which, if complied with by the public, will afterwards be transmitted to the house of commons. . . . Permit me then to remind you to be upon your guard, consider coolly what you are about, think of men and times past, examine whence this petition comes recommended, and you'll find it to be at the appointment (perhaps the modest request, or rather, the awful and powerful sanction) of the truly honourable body of Manchester Navigators.
>
> I might have spared the compliment, for trees are known by their fruits, and none can be mistaken in their judgements of the uprightness of these gentlemen's intentions; their past bewrays them, the facts are recent, witness their attempt when Sankey-wharf was first erected, their stoppage of vessels and

[1] The map on page 18 shows the direct competition that the new route would provide.

The mania years: by the 1790s the ewspapers were full of canal dvertisements – canals wanting ioney, canals wanting men, canal ieetings, canal shares for sale. This xample is from the *Leeds ttelligencer*.

or of Mr. Wm. Johnſon, at the Mills.

MASONS, STONE-GETTERS, & DIGGERS.
Leeds *and* Liverpool Canal.

TO be LETT, on Wedneſday the 16th Day of March Inſt. at Ten in the Forenoon, at the Black-Horſe, in Skipton, in the County of York,

The BUILDING of One LOCK, One AQUEDUCT, and Two PUBLIC BRIDGES, in, under and over the above Canal, in the Townſhips of Gargrave and Thorleby.

Alſo, The GETTING of STONES at Brumfitt-Cragg, near Silſden, proper for Hollow Poſts, Sills, Capings and Throughs, for Locks.—The Dimenſions may be had by applying to John Hammond, at Bingley Lock.

ALSO, *To be LETT*,

At the Houſe of Mrs. Hargreave, the Red-Lion, in Colne, in the County of Lancaſter, on Friday the 18th Day of March Inſt. at Ten in the Forenoon,

The BUILDING of Eight LOCKS, and ſeveral PUBLIC and OCCUPATION BRIDGES, at the Weſt End of the Summit of the above Canal.

Alſo, The CUTTING and COVERING of a certain Part of the TUNNEL in the above-mentioned Canal, which is intended to be laid open, and then arched with Maſonry.

Alſo, The MASONRY of the above-mentioned Part of the ſaid Tunnel.

☞ The Plans and Specifications of the above-mentioned Locks, Bridges and Aqueduct, may be ſeen, and all further Particulars had, by applying to Mr Whitworth, the Engineer upon the above Canal; or to Mr. Matthew Oddie, both at Colne aforeſaid; or to Mr. James Fletcher, in Bradford, in the ſaid County of York.

5 A Monmouthshire canal share from the 1790s, which specifies that the Company will also be making 'Rail Ways or Stone Roads'.

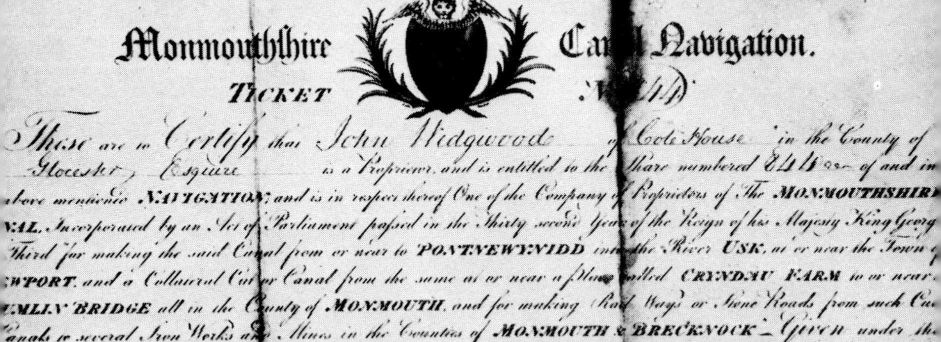

Monmouthſhire Ca[...] Navigation.

TICKET [...]

These are to Certify that John Wedgwood of Cote House in the County of Glocester Esquire is a Proprietor, and is entitled to the Share numbered 844 of and in [the a]bove mentioned NAVIGATION; and is in respect thereof One of the Company of Proprietors of The MONMOUTHSHIRE [CA]NAL, Incorporated by an Act of Parliament passed in the Thirty second Year of the Reign of his Majesty King George [the] Third for making the said Canal from or near to PONTNEWYNIDD into the River USK, at or near the Town [of NE]WPORT, and a Collateral Cut or Canal from the same at or near a Place called CRYNDAU FARM to or near [CR]UMLIN BRIDGE all in the County of MONMOUTH, and for making Rail Ways or Stone Roads from such Ca[nal or C]anals to several Iron Works and Mines in the Counties of MONMOUTH & BRECKNOCK — Given under the [Com]mon Seal of the said Company at their General Assembly at NEWPORT, the Twenty Ninth day of April in [the] Year of our Lord One Thousand, Seven Hundred a[nd] Ninety Nine.

[T]his Share may be Sold and Transferred [by a]n Instrument set forth in the Act.

A Memorial of every Transfer is to be Entered [in a] Book, and a Certificate thereof endorsed on the [Inst]rument, by the Company's Clerk.

And no Purchaser can receive any Dividends or Profits thereon, or Vote as a Proprietor thereof, until such Memorial hath been so entered.

All Shares are to be deemed part of the personal Estate of the respective Owners.

[signature] Clerk

6 Cruikshank turned his eye on the canal mania. Here the promoters puzzle over the engineer's plans – or go to sleep over his report.

7 'A Lack Water Canal'. Cruikshank's suggested solution for overcoming the objections of mill owners.

goods on the most base and frivolous pretences, their long and extravagant freights, tonnage, wharfage etc.[1]

The Mersey and Irwell Company lost their fight with the Duke of Bridgewater, only to find themselves involved again when the proposals for the Trent and Mersey canal were published. This threatened an even greater erosion of their trade, and they were joined in their opposition by a second navigation company, the Weaver. The pattern of the previous fight was repeated: initial opposition, a panicky attempt at compromise, more opposition. Their basic argument was that they were being treated unfairly – the navigations' proprietors had put a great deal of money into their works on the understanding that they would enjoy a trading monopoly. Now, they complained, they were faced by a rival scheme that would 'involve two effectual Navigations, established *under the Faith* of Parliament, in certain and irreparable Ruin'. Having delivered themselves of the main argument, they then went on with a general villification of the Trent and Mersey promoters, ending with a plea that the Bill approving the Trent and Mersey canal should at least be delayed:

> [The canal] could only be desired for *private* views, or to make a *more lucrative* Job for Engineers . . . an eager Solicitude for carrying this great and complicated Bill, in all Events in this Session, must flow from a Jealousy, that the scheme would be *generally* disapproved, and the Subscribers *revolt*, if a public Opportunity were given of examining and discussing the full and real Tendency of it, with an Attention and Severity equal to its Importance.[2]

The Trent and Mersey replied with pamphlets of their own, which failed to impress the Navigation proprietors: 'Two miserable Pieces . . . having been cautiously circulated, the *true Friends* of the Scheme for "opening a Communication between

[1] Pamphlet by John Hart, directed to 'The Gentlemen and Tradesmen at Warrington', quoted in Anon., *A History of Inland Navigations* (1779).
[2] A pamphlet of April 1766 entitled 'Seasonable Considerations on a Navigable Canal Intended to be cut from the River Trent . . . to the River Mersey'.

Hull and *Liverpool*"chearfully embrace a second Opportunity, of exposing to public View the flimsy Pretexts, under which the Enemies of Justice are driven to shelter themselves.'[1]

The river navigations continued to raise what support they could in advance of the argument being finally settled by Parliament. Ironically the main argument of the navigations – unfairness – was borrowed by the canal companies themselves when threatened by rivals and, at a much later date, when faced by the railways.

It was inevitable that the success of the first canal schemes would lead to a rush of new promotions, many of which would be in conflict. The fight between rival canal companies was every bit as vituperative as the earlier arguments had been. W. Hutton, in his *History of Birmingham*, gives a full account of one of the most fiercely fought battles, over the Bilston Canal Act, which is worth quoting at length. The original company is the Birmingham Canal Company:

> The growing profits of our canal company, already mentioned, had increased the shares from 140*l.* in 1768 to 400 guineas in 1782. These emoluments being thought enormous, a rival company sprung up, which, in 1783, petitioned Parliament to partake of those emoluments, by opening a parallel cut from some of the neighbouring coal-pits, to proceed along the lower level, and terminate in Digbeth.
>
> A stranger might ask, 'How the water in our upland country, which had never supplied one canal, could supply two? Whether the second canal was not likely to rob the first? Whether one able canal is not preferable to two lame ones? If a man sells me an article cheaper than I can purchase it elsewhere, whether it is of consequence to me what are his profits? And whether two companies in rivalship would destroy that harmony which has long subsisted in Birmingham?'
>
> The new company urged 'The necessity of another canal, lest the old should not perform the benefits of the town; that twenty per cent. are unreasonable returns: that the south

[1] Supplement to 'Seasonable Considerations. . . .'

country teams would procure a readier supply from Digbeth than the other present wharf, and not passing through the streets, would be prevented from injuring the pavement; and that the goods from the Trent would come to their wharf by a run of eighteen miles nearer than to the other'.

The old company alleged, 'that they ventured their property in an uncertain pursuit, which had it not succeeded, would have ruined many individuals; therefore the present gains were only a recompense for former hazard; that this property was expended upon the faith of Parliament, who were obliged in honour to protect it, otherwise no man would risk his fortune upon a public undertaking; for should they allow a second canal, why not a third; which would become a wanton destruction of right, without benefit; that although the profit of the original subscribers might seem large, those subscribers are but few; many have bought at a subsequent price, which barely pays common interest, and this is all their support: therefore a reduction would be barbarous on one side, and sensibly felt on the other; and, as the present canal amply supplies the town and country, it would be ridiculous to cut away good land to make another, which would ruin both'. . . .

Both parties beat up for volunteers in the town, to strengthen their forces; from words of acrimony, they came to those of virulence; then the powerful batteries of hand-bills and news-papers were opened; every town within fifty miles, interested on either side, was moved to petition, and both prepared for a grand attack, confident of victory.

The new promoters actually had rather more force on their side than the partisan Hutton suggests, for, as well as providing the parallel cut which caused all the fuss, they were also proposing to build a link between the Coventry and Stafford canals, which would certainly be a very useful waterway. Eventually a compromise solution was reached by which the Birmingham Company agreed to build the Stafford-Coventry link and the rivals withdrew. Hutton summed up the affair: 'Thus the new proprietors, by losing, will save 50,000*l.* and the old, by winning, become sufferers.' He forgot to mention that, for once, the public won.

Some canal promoters became so deeply involved in arguing with each other that the original point of the argument tended to get lost, and they finished up with neither party doing anything. A meeting was held at Wells in Somerset on 5 December 1792 to discuss a proposal for a canal from Bristol to Taunton. A week later a rival set of promoters was advertising their public meeting:

> They are aware that a Meeting has already been held at *Wells,* for the purpose of carrying a Canal from *Taunton to Bristol,* but not thinking themselves candidly dealt with, by the manner in which the meeting was convened, both from the unusual shortage of the notice given (*being only one clear day*) and the studious concealment of its object; they conceive that no resolutions then formed, nor measures then concerted, should preclude their adopting and endeavouring to carry into execution a Scheme replete with equal, and perhaps superior, advantages to the Town of Taunton and Country adjacent.[1]

They were still talking almost twenty years later.

When they were not fighting each other, the canal promoters had other problems to contend with, notably objections from landowners. By no means all of the aristocracy and the larger property owners shared Earl Gower's view of the identity of interest between the trading and landed classes. Many of the landowners, the country squires, were opposed to canal schemes. Some saw the canals as an irritating intrusion, a disturbance of their privacy:

> The Petitioners, being Owners or Occupiers of Houses and Lands near to or through which the said Canal is intended to be cut, apprehend that great Injury will arise to them . . . by the near Approach of the said intended Canal to the Houses and Pleasure Grounds of several of the Petitioners, which they have made at great Expence.[2]

[1] *Felix Farley's Bristol Journal,* 15 December 1792.
[2] *Journal of the House of Commons,* 23 January 1783.

But others saw the canals as a threat to something much more important than the view at the bottom of the garden. The promoters of the Ashby-de-la-Zouch canal became involved in a long and bitter argument with a Mr Curzon who owned a large house and grounds near the proposed route of the canal. At first, the argument seemed to be a common enough sort of thing – Curzon complained that the canal would cut him off from the source of water that supplied his house. The canal promoters were used to this sort of problem, and offered various solutions which would guarantee a continuance of the supply. But Curzon refused to listen to any suggestions put to him, and the argument moved from private debate to public quarrel. In their final offer the canal promoters proposed laying pipes from the spring to bring water right to his door, 'Which he is now obliged to fetch in water-carts at the distance of *half-a-mile* from his house'. Still, Curzon refused the offer and the promoters, in final exasperation, brought the argument back to a more fundamental level:

> As Mr Curzon has recourse to interrogatories he will do well to answer the following: Did he not state, as the original ground of his opposition, his objection to the extension of Trade and Manufacture in the country? To be more explicit, did he not, in a conversation on the subject, declare his apprehensions that, if the projected canal should take place, Ashby and Measham would become places of *as great trade as Manchester and Sheffield*?[1]

This was the real core of the argument: the water-supply was simply a pretext. And, in his own terms, Curzon was absolutely right. The canals, the increased trade that they would bring, the spread of manufacturing – these were all forces that threatened his world and his way of life. Above all, they were forces that threatened his political influence and his power. His fight with the Ashby-de-la-Zouch company was not a fight about water supply, but a fight against the whole progress of the Industrial Revolution, of which the canal system formed a vital part.

Not many of the opponents of canal systems saw things as

[1] *The Times*, 16 March 1793.

clearly as Mr Curzon appears to have done, but he had many supporters for his view. What is perhaps surprising is that he did not have more. Even the Ashby canal opponents were in a minority: along the 50-mile length 'excepting himself and one other person, there are not more than 8 miles of land-owners dissenting, notwithstanding the most *unusual* exertions have been made by *interest* and *misrepresentation*, to extort dissents'. Other canal companies fared even better: the promoters of the Basingstoke canal, for instance, found 'but two dissenting voices' along the 43-mile route. The millowners, the other main class of landowner concerned in the opposition, were not as a rule, against canals as such: they simply objected to any scheme that threatened their water supplies. They were, however, concerned as much with receiving adequate protection for their own interests as with direct opposition.

Any canal promotion was always liable to run up against a variety of local interests or individual objections. When the Birmingham to Warwick canal was proposed, there was immediate opposition in Birmingham from a group who were worried in case the new canal would be used to take away coal from the pits that normally supplied Birmingham itself, leading to a coal shortage and higher prices. This kind of specific economic argument was common. Less common was the view put forward by a gentleman objecting to the Regent's canal. He had a surprisingly modern opinion to express:

> Nature appears to afford to particular spots conveniences for trade; to others, pleasures derivable from local beauties. The intermixture of these is frequently heterogenous, notwithstanding every effort to render them otherwise. A steam-engine at Limehouse, or Bankside, is hardly objectionable; in St James's or Hyde-park, it would be intolerable. Is there, then, no part of the environs of London which shall escape the innovations of speculators, builders, water-work and canal schemes?[1]

Finally, like any enterprise that attracts public scrutiny, a

[1] *The Times*, 17 December 1811.

canal promotion also attracted its fair share of cranks, who had their own theories on the evils of a canal system: Mr J. H. Pugh, observing that the English climate was worsening, felt that he had an explanation:

> An idea has suggested itself to the author, that although our climate, from our situation in the hemisphere, must of necessity be extremely variable, yet that the moisture and concomitant evils may be increased by the frequent and extensive canals and aqueducts that are made for the facilitation of commerce.[1]

And yet more ammunition for objectors was provided by the shocking use to which a canal might be put:

> Gentlemen, – I take this opportunity of informing you, that the neighbourhood is exceedingly annoyed by the practice of public bathing in your canal; a practice which is carried to a very great extent. I am confident it is unnecessary, either to point out the indecency of such a practice, and particularly its extreme offensiveness to females in so public a situation, or to urge any reasons to induce your interference for its suppression.[2]

The canal promoters themselves were every bit as busy producing pamphlets and writing to the press as were their opponents. By no means all of their time was taken up with refuting arguments and allegations; they were also active in pushing forward their own arguments for the canals. They went about it with enthusiasm and panache.

The main theme, naturally enough, in all pro-canal writings is the economic one, and, of the different economic factors mentioned, coal predominates. The lower cost of coal, the increased supply of coal – the same point is banged home again and again by the pamphleteers and letter writers:

[1] *Gentleman's Magazine*, 1795.
[2] Letter to the Proprietors of the Grand Surrey Canal, *The Times*, 12 September 1811.

> I beg to ask the opposers of the navigation from Coventry to Oxford, whether they think all the inconveniences they have pointed out, are equal to the starving to death many hundreds of poor inhabitants in the inland countries, where coals may be brought by such a navigation? For I do take upon me to say, that were the overseers of the poor ever so much disposed to show compassion to the poor, they have it not now in their power to procure them fuel: There are many places near Banbury, Brackley, Bicester, etc where coals cannot possibly be had at any price to supply the necessary demand for them . . . there are no materials by which the poor can procure even a wretched fire. In these parts, I believe, they often perish by cold. . . . And they must not have fire, forsooth, to keep them alive, for fear that Cheshire cheese should go to London by way of Oxford, or coals should be raised 1d a hundred in Warwickshire![1]

Carriers and colliers joined in to reinforce the argument on reduced costs. Faced by 'the laboured attempts of the Opponents to deceive the Public as to the difference between Land and Water carriage', an Ashby carrier printed his guarantee of being able to reduce the price of coal at Hinckley so substantially that 'the Saving to that Town in this single article will be 2500 Pounds annually'.[2]

As well as coal, different commodities were advertised for different regions, and most pro-canal pamphlets reeled off long lists of items, all of which would be supplied more easily and more cheaply. The Trent and Mersey was advertised as being able to improve the trade in 'rock salt, stone, coal, limestone, alabaster, slate, manure, iron-ore, corn, timber, Cheshire cheeses, manufactured salt, earthenware, finished metal goods' and so on.

The pamphleteers also bore in mind Erasmus Darwin's advice to woo the landowners, and went into raptures over the delights that the fortunate few, lucky enough to have a house along a canal route, could expect:

[1] Anon., *A History of Inland Navigation* (1779).
[2] *Leicester Herald*, 12 January 1793.

> Having considered the principal advantages which the public may reasonably expect from the execution of this design, we ought not to forget the pleasures that may arise from it to individuals, especially as taste is so universally cultivated, that our farms are gradually improving into gardens. And here it must be allowed, that to have a lawn terminated by water, with objects passing and repassing upon it, is a finishing of all others the most desirable. And if we add the amusements of a pleasure-boat that may enable us incessantly to change the prospect, imagination can scarcely conceive the charming variety of such a landscape. Verdant lawns, waving fields of grains, pleasant groves, sequestered woods, winding streams, regular canals to different towns, orchards whose trees are bending beneath their fruit, large towns and pleasant villages, will all together present to the eye a grateful intermixture of objects, and feast the fancy with ideas equal to the most romantic illusions.[1]

The pamphleteers were also adept at producing new and ingenious arguments in favour of canals. At the time of corn shortage, they showed how the supply could be increased by the growth of canal transport: 'One horse by the canals will draw as much as sixty by the roads, and the provender for the fifty-nine horses extraordinary requires at least four hundred acres of land.'[2] This argument became a great favourite and was repeated many times, not always with quite such exaggerated ideas of the relative efficiency of land and water transport. A few, however, went even further: 'Last week 840 sacks of flour arrived by the Kennet and Avon Canal at Bath, from Newbury. . . Six horses only were employed on the above freightage,

[1] Phillips, *op. cit.* This is taken from Bentley's pamphlet for the Trent and Mersey. In an earlier version he had replaced the phrase 'terminated by water' to 'terminated by a canal', which had brought out the critic in Wedgwood: 'Why change a more elegant & equally simple word for a worse? Why a Canal is as straight as *Fleet Ditch* – A Canal at the bottom of your meadow! Foh! It can't be born by the Goddess of modern taste, but "*Water*" ay *Water* give me *Water* to terminate or divide my Lawn.'

[2] *Morning Chronicle*, 11 January 1793.

which by land carriage would have required one hundred and ninety.'[1]

For over half a century, the handbills and pamphlets came streaming out from the presses, arguing for and against canal promotions. All had the same object: to raise local and national support in the battle that really mattered – the battle to get a Canal Act through Parliament.

[1] *Shrewsbury Chronicle*, 8 February 1811.

5

TO PARLIAMENT

It is the unanimous opinion of this assembly, that the success of the Act for establishing this Navigation hath been greatly owing to the warm and effectual support derived from the Right Honble. the Earl of Hertford and his family in the several stages of the Bill through both Houses of Parliament.
Coventry Canal Minute Book, 19 February 1768.

No matter how energetic the industrialist might be in campaigning for canals, or how enthusiastically the pamphleteer might sing out the praises of navigation, when it came to the crucial stage of getting the Bill through Parliament, there was no substitute for the 'warm and effectual' support of a Right Honourable Member.

Parliament, throughout the eighteenth century, was a small, enclosed world with a membership drawn from a very narrow social band – to achieve any kind of success, it was necessary to know someone of influence, someone who could preferably call on a wide range of family connections within the two houses. Canal promoters, such as Josiah Wedgwood, might exert influence within their own communities, but they were without any direct access to political power. This was still the age when a Member of Parliament sat for what William Cobbett called the 'accursed hill' of Old Sarum, while a thriving manufacturing town like Birmingham was without any representation at all. So the business of obtaining a Canal Act was partly a matter of arguing a case, and partly a matter of obtaining the right sort of support.

The Duke of Bridgewater, as a member of the aristocracy himself, had less of a problem than some of his successors. But, as the first in the line of canal promoters, he found that he had a great deal of explanation to provide on the general principles and practices of canal construction. Before being brought to a vote in the two Houses, the plan for the canal extension to the Mersey had to be explained in detail to the members of the special Committee of the House of Commons set up to examine the proposals. James Brindley, the Duke's engineer, was called in to provide the answers to the technical questions. Very few Committee sessions in the House of Commons can have been livelier than those involving James Brindley and his illustrated lectures on canals.

Brindley was no scholar, in fact he was almost illiterate, but he had a practical man's gift for practical exposition. If the Committee wanted to know how a canal could be made watertight, then to Brindley the only sensible thing to do was to send out for clay and water and give a practical demonstration of the art of 'puddling'. When a bridge had to be described, Brindley preferred a demonstration that involved carving up a cheese with his pocket knife to oral reasoning. Whenever he got stuck for words, he would take out his piece of chalk, get down on his hands and knees, and draw diagrams on the floor of the committee room. It was apparently a popular saying in Lancashire at that time that 'Brindley and chalk would go through the world'.[1] He certainly went through the Parliamentary Committee with a rare abandon.

The opposition to the Duke's plans was headed in Parliament by Lord Strange, a Lancashire man who acted as spokesman for the Navigation interests. He was a Tory, and the opposing forces tended to split along party lines, the Tories supporting Lord Strange, the Whigs usually supporting the Duke of Bridgewater. Brindley followed the debate with keen interest, and recorded his impressions of its progress. At first he felt gloomy and apprehensive as the opposition case was put; 'The Toores mad had agane ye Duk' [the Tories made head against the Duke], but he

[1] Samuel Smiles, *Lives of the Engineers*, Vol. 1 (1874).

became more optimistic as the petitions supporting the Duke began to flood into Parliament and, finally, he was able to triumphantly record, 'Lord Strange defetted!' The Bill received the Royal Assent on 4 March 1762.

The Duke of Bridgewater's and the other early canal bills were infrequent interruptions to Parliamentary business – trivial matters compared with the problems of international affairs, particularly when a minor disturbance in the colonies suddenly turned itself into the American War of Independence. But for the protagonists, canal bills were major preoccupations. When Parliament was petitioned to pass the Trent and Mersey Bill, Wedgwood had to write to his friend Bentley:

> Pray put your business into such a channel, as it will move along with the assistance of your good partner for a few months, for I am afraid in that time you will not be able to follow it much yourself.[1]

In fact, the Bill did not take quite that long, but some canal bills did, and the promoters always had to face the prospect of a series of deferments.

When the American War of Independence ended and commercial confidence began to be restored following the collapse of the disastrous administration of Lord North in 1783 and its replacement in 1784 by a new Government headed by the young William Pitt, Parliament, in its turn, found canal business occupying more and more of its time. By the 1790s canal enthusiasm had grown into canal mania: 'One Hon. Member wished his grand-children might be born webb-footed that they might be able to swim in water, for there would not be a bit of dry land in this island to walk upon.'[2]

Faced by the growing number of canal bills being presented, Parliament tried to bring some order to the chaotic scene by means of a special set of Standing Orders, which were introduced in 1792. The Orders laid down that all canal proposals should be

[1] Eliza Meteyard, *The Life of Josiah Wedgwood* (1865).
[2] *The Times*, 2 March 1793.

advertised in the *London Gazette* and local papers before presentation to Parliament, that a map should be deposited with the local Clerk of the Peace, that proper costings should be worked out, and that lists of landlords, assenting and dissenting, who had property along the proposed route, should be made. It was also ordered, somewhat optimistically, that bills should contain provisions for ensuring that subscribers paid up on demand. Even with the new, stricter, rules the number of canal schemes continued to grow. 1793 was the peak year. At the beginning of the year, a writer in *The Times* wrote that: 'The Canal Bills so multiply, and the petitions for and against them so encrease, that they promise to be as tedious a business as the Trial of Mr Hastings.'[1] By March, there were no less than thirty-six canal bills under the consideration of Parliament, and nineteen received the Royal Assent during the year. Inevitably, there was a drop in subsequent years, and by the time the Napoleonic Wars had begun the number had fallen right away.

The discussion of the canal bills of the 1790s was a very different matter from the chalk and cheese lectures of Brindley. Arguments were taken away from the laymen and handed over to the lawyers, as all parties became more expert and the arguments more complex. Where a few witnesses used to be called, they now came in crowds – *The Times* reported that almost two hundred witnesses were brought up to London for the Birmingham Canal Bill of 1791. All those petitions that had been so assiduously canvassed by the promoters and opponents came thudding down on to the table in Parliament by the score. And it was not just the number of petitions that increased:

> A petition from Birmingham, in favour of the Worcester and Birmingham canal, was yesterday presented by Sir George Shuckburgh to the House of Commons, which, from its singularity, deserves notice; being signed by 4058 persons, and the petition measured in length above fourteen yeards.[1]

[1] *The Times*, 16 February 1793. Mr Hastings was Warren Hastings, whose trial had been keeping Parliament busy for months.
[2] *Leeds Intelligencer*, 15 March 1791.

When an important argument blew up, as in the case of the Bilston Canal Act, the unfortunate Members who formed the Committee set up to study the proposals almost disappeared under the avalanche of papers. Petitions from local towns, both for and against, were presented, local big-wigs sent their views of the matter, local tradesmen sent theirs. Some petitioners caused confusion by changing their minds, and sending in two petitions, as did the citizens of Stafford. The proprietors of the Birmingham canal put the finishing touch to their pleas of the hardship that would be caused to their subscribers, by a tear-jerking document from a selection of the shareholders headed 'A Petition of many Widows and Orphans'.[1]

The increased professionalism of the advocates meant that each canal bill tended to take up more of the House's time than ever. When the Dudley Bill went into Committee in the House of Lords, it was debated for twenty-three days, with a division almost every day. When the debate reached the floor of the House, it attracted a great deal of attention – 'This Dudley Canal Bill has some secret charm to draw their Lordship's attention'. It was also the scene of a rather forlorn rearguard action by a representative of the diminishing minority who still opposed the whole concept of canal navigation. Lord Portchester 'entered into a long dissertation of its demerits. . . . He combated the generality of Canal Bills, considering them as a very great evil, and compared them to an act of national justice, such as a war of public necessity, wherein a number of privateers took advantage of the occasion and attacked private property.'[2]

But however important these debates might be, very little reached the pages of the newspapers. *The Times* comment was typical: 'A long debate took place on the Dudley Canal Bill, which lasted from four until nearly eight o'clock. The arguments *pro* and *con* would be uninteresting to the public, and therefore we shall not detail them.'

In one sense they were right, for the most important and interesting part of the Parliamentary battle took place outside the

[1] *Journal of the House of Commons*, 31 January 1783.
[2] *The Times*, 1 May 1793.

walls of Westminster. This was the fight to win over the vital interests to one side or the other – and it was fought by accountants rather than lawyers.

During the end of 1792 and early 1793, the Lancaster Canal Company became engaged in a feud with the Leeds and Liverpool Canal Company. Basically, the argument arose because the Leeds and Liverpool Company wanted to deviate from their approved line, but this deviation would have brought their canal into the area set aside for the southern end of the Lancaster. The specific details of the quarrel are less important than the lines along which it developed.

The first sign of dirty dealings was spotted by the Lancaster Company in August 1792:

> The Committee have been informed from Lord Balcarres & others, it is industriously propagated that the Line on the South Side of Ribble is not meant to be carried into execution.
>
> We apprehend these reports are spread by the Agents of the Leeds and Liverpool Canal in order to gain the assent of the Land Owners to their projected Deviation.[1]

They sent off a warning note to 'their' Member of Parliament:

> The Canal Come take the liberty of requesting you will desire your friends in Parliament not to be too hasty in engaging their interest to the Leeds & Liverpool Proprietors, as we do not yet know the effect their scheme may have upon our Canal.[2]

The Lancaster Canal next became interested in the activities of a Mr E. Gorst who had been heard to refer to them as 'narrow minded men and blind to their own interest', so that when they also heard that Mr Gorst was travelling around their part of Lancashire, collecting signatures for a petition for 'a general junction of the Canals in this Country' they concluded that 'Mr Gorst under a false pretence of gaining a Petition for a General

[1] *Lancaster Canal Letter Book*, 22 August 1792.
[2] *Ibid.*, 4 November 1792.

Junction is merely serving the interest of the Bolton extension of the Leeds & Lpool.'[1]

Things were clearly moving towards a direct confrontation of the two parties, and the Lancaster Proprietors began to organize their Parliamentary forces. They began by reassuring their more important supporters within Parliament: a letter was sent to the Earl of Derby refuting 'the Bugbear held out by the L & L to alarm our proprietors and poison the minds of individuals who were our friends when we Solicited our Bill last Sessions'.[2] The Company's Solicitor, Mr Jackson Mason, was sent down to London to keep an eye on things, but he was a gloomy man and soon found that Parliamentary intrigues depressed him: he was, he reported to the Lancaster Committee, 'in the cellar'. He even began to have doubts about the loyalty of Lord Balcarres, who was one of the leading spokesmen for the Lancaster interest. The Committee knew how to cope with that:

> I do not see why you should doubt his Lordship's Integrity – but if they should propose something very advantageous to his Lordship & you should find him *begin to Waver*, it may be well to drop a hint to him that should he permit the L & L to succeed it may be necessary for us to consider whether a lower level could not be carried.[3]

This last threat was quite simply economic blackmail. The Lancaster Company had an agreement with Lord Balcarres to drain his extensive coal mines in the area to supply the water for the canal – they were now reminding his Lordship that such an advantageous scheme was, of course, dependent on his Lordship's continuing political support. The 'warm and effectual' support of Right Honourable Members had to be kept up to the mark. On this occasion, the Lancaster Canal's supporters were successful in getting the Leeds and Liverpool plans thrown out, but the next year they were back, and the political and financial bargaining began all over again. One important landowner with political

[1] *Ibid.*, 30 December 1792.
[2] *Ibid.*, 3 February 1793.
[3] *Ibid.*, 2 May 1793.

E

influence could be relied on to support the Lancaster Company because, as they reminded him, he could expect better coal sales from their canal that 'goes thro' a Country of a great extent that has not a coal within itself'.[1] Another would support the Leeds and Liverpool: 'Lord Petre is most certainly deeply interested in their success, and exerts himself much. He told me last Spring that it would be 8 or 900 a year in his way if they succeeded.'[2]

As well as indulging in what, to modern eyes, seem extraordinarily open exercises in bribery and corruption, the two Companies also engaged in a variety of purely political ploys and manœuvres. The Lancaster Company, for example, proposed a whole set of new clauses for the Leeds and Liverpool Bill – clauses which they knew were completely unacceptable, and which were designed purely to waste the opposition's time and, hopefully, to antagonize some of the Leeds and Liverpool supporters, who might be fooled into thinking that the Lancaster company were trying to be helpful. Then, in March 1794, the two Companies gave up their bickering, agreed on a compromise solution, and, after two years of insults and double-dealing, declared themselves the best of friends.

All this juggling of political and financial interests was exceedingly complicated, and both Parliamentarians and Canal Proprietors had to be very careful not to upset the delicate balances. When the Grand Junction Canal Company was going to Parliament, they naturally sought the support of other canal companies, including the Lancaster. Since these two canals were at opposite ends of the country, there was no possibility of any conflict of economic interest. But, before agreeing, the Lancaster Committee had to check with their Parliamentary supporters to see how such a move would fit into the political pattern:

> The Committee have received a letter from the Grand Junction Canal to send petitions from their Comp., the Corporation and the Town in their favor – we wish to know your Sentiments of the Grand Junction Interest, and if our sending

[1] *Ibid.*, 15 January 1794.
[2] *Ibid.*, 24 February 1794.

petitions in their favor will in any wise benefit them, or if any of our friends are in opposition to them.[1]

Even when economic considerations were plain, a company might have to modify its plans if it appeared to be politically expedient. In an earlier conflict between the Lancaster and the Leeds and Liverpool Companies, the Leeds proprietors were raising objections to the former's plans. They received private information, however, that their opposition was putting them in bad odour with the Speaker of the House, who was against it 'in principle' and considered the particular form of the opposition to be 'without precedent'.[2] The Leeds and Liverpool took the hint and dropped out of the contest. As it turned out, things went well for the Leeds and Liverpool on this occasion, as their opposition had the effect of panicking the Lancaster Proprietors, who rushed their Bill forward with 'many imperfections', and the Leeds and Liverpool found themselves with legitimate opportunities for opposition.

It was not only the canal companies who were able to exert pressure. Local interests could gain considerable concessions in payment for their political support:

> The water communication with London and Bristol effected by means of the Canal in question [the Kennet and Avon] has proved to be the occasion of no small advantage to the commerce of Devizes. A very considerable additional outlay in the cost of that undertaking was incurred by bringing it so close to this town, owing to the rapidity of the ascent which it is made to surmount in order to gain the high ground just at this particular spot; whereas the attainment of the same level might have been accomplished by pursuing a line of much more gradual aclivity. Whatever may have turned out to be the result of this arrangement, whether to the interest of the shareholders on the one hand, or those of Devizes on the other, it is not generally known that the Corporation of the town, through

[1] *Ibid.*, 18 December 1792.

[2] *Leeds and Liverpool Canal: Leeds Committee Minute Book,* 1 June 1792.

the influence of their representatives in Parliament, were the main instruments of bringing it about.[1]

The detour, in fact, involved the Company in the construction of one of the most spectacular flights of locks in Britain – a feature which has brought more pleasure to later canal enthusiasts than it did to the Kennet and Avon's shareholders.

The Parliamentary intrigues of the eighteenth century would have taxed the ingenuity of a Machiavelli, so it is hardly surprising that many canal bills were never passed, and that many others only got through on the second or third attempt. The acts that were finally agreed reflect, in their various clauses, the conflict of interests that preceded their acceptance.

The main part of any canal act was concerned with defining the line the canal was to take (usually referred to as 'the Parliamentary line', deviations from which required separate authorization from Parliament), confirming the estimated costs of construction and authorizing the necessary capital, and regulating the tolls and tonnage to be charged when the canal was completed. Procedures were also laid down for the purchase of land along the route, including the compulsory purchase powers of the canal proprietors, and the composition of the various bodies that had to be set up to adjudicate between the different parties where agreement could not be reached. These main clauses became fairly standardized as time went on – it is the special clauses, introduced to define the position on some particular issue, that show where the real hard bargaining had gone on.

The mill owners were one group who sent in petitions objecting to virtually every canal scheme that came before Parliament, and always in much the same terms. The canal would bring ruin on them by 'intercepting several Streams and Currents of Water, which at present communicate with and run into, the River Thame, whereby the Water (already insufficient for the supply and working of several Mills upon the said River) will be con-

[1] James Waylen, *Chronicles of the Devizes* (1839).

siderably diminished'.[1] Substitute the name of any river for 'Thame' and you have the all-purpose millowners' petition: canals, to them, were always a menace, just as their own water supplies were always already insufficient. Sometimes their opposition was so effective that they could get a canal bill held up for years. The Rochdale Canal plans were first put before Parliament in 1791, but the bill was delayed until 1794 and then was passed only when the Company agreed to build special reservoirs to protect the water supply to the mills. In most cases, however, the mill owners were satisfied with getting suitable safeguards written into the act, such as the clause in the Trent and Mersey Navigation Bill 'for preventing injury or damage to the Owners and Occupiers of Mills upon the River Churnett, and other Persons interested in the Water of the said River, or the Streams and Waters flowing into the same'.

Sometimes specific properties were protected: Mr X's orchard must not be disturbed; Mr Y's view must be unobstructed. Farmers along a canal route would often receive special benefits, apart from the benefit of irrigation: many bills, for example, contained provision for manure to be carried on the canal free of tonnage.

The job of getting a canal act passed, of reconciling opposing interests, was a long and tedious one. It was also expensive. There was usually at least one full-time Company man employed in Westminster to marshal the forces, there were the lawyers fees to be met, witnesses' expenses to be paid and palms to be greased (the Duke of Bridgewater's accounts, for example, contain suspiciously vague entries headed 'Mr Bill'). It is not surprising that after the expenditure of so much time, effort and hard cash, the actual passing of a bill should be an occasion for general celebrations. Local poets did their worst and there was a great deal of revelry and mutual congratulations. When the Trent and Mersey Act was approved, it was marked by a ceremony on 26 July 1766 in which Josiah Wedgwood cut the first sod of earth, after which there were 'some excellent speeches, in which, like

[1] *Journal of the House of Commons*, 23 January 1783.

veterans, the promoters of the scheme recounted their toils and dangers in the battle won'.[1]

The veterans had earned their congratulations. But, having got authority to build a canal ,they still had to find the money to pay for it.

[1] Meteyard, *op. cit.*

6

RAISING THE MONEY

> At a late meeting of subscribers to the Rochdale Canal, 53,000*l.* was immediately raised . . . and so unanimous are the Gentlemen in the neighbourhood for carrying it into effect, that three times the sum can be forthcoming whenever it is necessary.
>
> *The Times*, 4 January 1791.

There was a time in the history of canal building when it seemed that a canal promoter had only to announce his plans to the world for the money magically to appear. It was not always like that, and even projects that began with such golden promises of unlimited finance sometimes ended with those promises unfulfilled. The despairing complaint of a Committee member of the Lancaster Canal could have been repeated at canal workings all over Britain by the end of the eighteenth century – 'Money, money, money is the common cry'.

After the Duke of Bridgewater's struggles to finance his own canal, the majority of the later works were constructed by Canal Companies, financed by shareholders. At first, the leading proponents of a canal were also its main backers. The Trent and Mersey Company raised their capital of £130,000 in £200 shares: the Wedgwood family held £6,000 worth between them, and less influential potters also subscribed; Earl Gower, who had been such an active supporter, had £2,000 invested, and even Thomas Gilbert, who had previously had doubts about this method of running a Company, came in with his £2,000 worth; the Duke of Bridgewater acknowledged the mutual interest of the Trent and Mersey and his own undertakings, and he too came in with £2,000; James Brindley, showing a commendable confidence in his own abilities as chief engineer to the project, put up £2,000, and his brother displayed confidence by doing the same. Naturally enough, with shares at £200 each, the backers were all substantial and wealthy men, and men with a direct interest in the success of the venture. The potters, for example, who between them raised £15,000, were more concerned in investing their money for the improvement of their trade than with investing for the possible return on their capital in terms of dividends paid, though no doubt they were glad enough to see the dividends when they came.

This type of subscription could be repeated many times in the early years of canal promotion. In 1767, the Birmingham Company raised its capital of approximately £70,000 by selling shares at £140, and by limiting each subscriber to ten shares. A canal such as the Huddersfield, built to serve special local interests, could rely on local support: the share book shows quite clearly that many of the subscribers were local men, manufacturers and merchants from Lancashire and Yorkshire – clothiers, cardmakers, woolstaplers, clothdressers, these are the descriptions that recur. Often the individual holdings were small, and it is quite common to find small traders having only one or perhaps two shares.

These local backers were supporters of trade, not financial speculators. But the remarkable profitability of many of the first canals brought about a change in attitude. The fortune being made by the Duke of Bridgewater began to be repeated elsewhere. Even in the years of trade recession, during the American War of Independence, the dividends on canal shares rose, and with the peace they rocketed. As the dividends went up, so did the share prices: the £140 Birmingham shares of 1767 had gone up to £370 by 1782 and ten years after that the price had shot up to £1,170! Not surprisingly, the pattern of investment began to change. At first, the growing interest of outside financiers was taken as a mark of the soundness of the investment. But soon, with the lure of such fat profits and juicy dividends, the gamblers and speculators moved in – the 1790s became the years of Canal Mania, foreshadowing the Railway Mania of the next century. New subscriptions were filled almost instantaneously, and brokers set up business in the large towns to deal with the growing trade in canal shares.

The climate was right for speculators – canal shares represented only one outlet for men with a little money who wanted to make a lot in a hurry: there were also the lotteries, the tontines, and gambling on anything and everything. The newspapers at the end of the century were full of accounts of the most bizarre wagers – bets on races of all sorts and descriptions, horse races or foot races, and there was even a race between a horse and a mule and the man who bet he could drink two glasses of lamp oil – anything seemed to be a possible starting point of a wager. The press reported them all with glee but were also ready to draw a moral if one was there for the drawing:

> Thursday se'nnight at a public-house in Bullock-Smithy, Manchester, a man for a trifling wager, drank three quarts of ale and three gills of rum – He fell a martyr to his impudence – he died immediately.[1]

Among all the other speculations, canal construction seemed as good a gamble as others, and safer than most: 'It is no wonder that the cutting of canals should be so favourite an object to this kingdom. . . . The Staffordshire canal is now said to pay the subscribers thirty per cent.'[2]

The scramble to get hold of canal shares really did rise to manic proportions. The speculators were not interested in examining the merits of a new scheme – any plan that had the magic word 'canal' in its title was sure of support; the canal share was a cornucopia that everyone wanted to get into their hands. Accounts of canal meetings no longer dwelt on the 'respectability' of the subscribers – the witnesses were too busy trying to work out what everybody at the meeting was worth:

1 *Leeds Intelligencer*, 9 July 1792.
2 *Ibid.*, 7 June 1791.

Upon a moderate estimate of the property of those present, it would amount to a sum of such immense magnitude, that we forbear making it, lest, to those unacquainted with their great opulence, our calculation might seem undeserving of credibility.[1]

Not everybody was blind to the dangers of thoughtless speculation. In a letter headed 'Navigation. Caution to the Public', one correspondent gave his views of a proposed canal from Burton to Fradley Heath:

From the great eagerness that has lately been shown to obtain Subscriptions in Navigation Projects, of every kind, it may be concluded that the Subscriptions to be opened at Wichner will be speedily filled, not by Persons who seriously mean to promote the undertaking, but by Speculators, who propose to traffic in the Sale of Shares, before the application to Parliament. For preventing in some degree this unwarrantable Commerce, and for guarding the unwary against its evil effects, the Public are now informed of the improbability of the above Scheme taking effect. If they will obstinately run into error after this caution, they will only have to repent, when too late, of their own folly and temerity.[2]

Not much notice was taken of such warnings and, in their eagerness to get in on a new subscription at the lowest price, speculators dashed wildly around the countryside from meeting to meeting. Once a rumour started that a new subscription was being opened, they would hire any broken-down carriage or aged horse and set off on the chase for the high dividend or the quick profit. The speculators bred impracticable canal schemes, just as the canal schemes bred imperceptive speculators.

So unbounded have the speculations in canals been, that neither hills nor dales, rocks nor mountains, could stop their progress, and whether the country afforded water to supply them, or mines and minerals to feed them with the tonnage, or

[1] *Felix Farley's Bristol Journal,* 3 November 1792.
[2] *Derby Mercury,* 1 November 1792.

whether it was populous or otherwise, all amounted to nothing, for in the end, they were all to be Bridgewater canals. His Grace's canal has operated upon the minds of canal speculators, much in the same manner as a large lottery prize does upon the minds of the inhabitants of a town....

It can be no matter of surprise, that the public mind should have been so much inflamed with canal speculations ... and caused many, who were determined to become canal subscribers, chearfully to submit to sleep in barns and stables, when beds could not be procured at the public houses, where meetings were held to receive subscriptions for newly projected canals.[1]

Not all the schemes that were begun during the mania years were ill-considered but, sensible or ludicrous, they were all lumped together in the public mind, and the long-term effect was to bring canal promotions into disrepute. Proposals began to appear for limiting the dividends on canal shares to discourage the wilder speculations and more blatant malpractices. At first, the matter was raised in the press:

Would it not be right that the interest of the money subscribed to these late projected canals should be limited by act of parliament? When I say limited, I do not mean to four or five per cent, but to ten or twelve. I do not pretend to say myself that it will ever amount to near this last sum; but I know that it has been said that the proprietors of the shares in some canals (particularly in the Grand Junction) will, in a few years after they are completed, obtain at least 25 or 30 per cent interest for their money. That men, who have the courage to risk 1000*l.* on a chance of success, should receive more than common interest, is certainly no more than justice; but when, at the very place of meeting, 20*l.* and 30*l.* were offered to any person who would only write his name, and transfer his right to the shares he subscribed for to another man (which was the case when the subscription was first opened for the Braunston canal), I cannot comprehend how it can be called an adventure

[1] John Sutcliffe, *A Treatise on Canals and Reservoirs* (1816).

> to subscribe. . . . Besides, canals are now become quite a lottery; and there is as much gambling going forward, to all appearance, on the buying and selling of shares, as in the Alley.[1]

A year later, the matter was moved out of the columns of the magazines and papers and into the House of Commons, when Mr Powys began a campaign for a limitation of the profits of canal 'adventurers'. In the event, legislation proved unnecessary. By the beginning of the nineteenth century, the outbreak of war and the trade recession had brought a reduction in canal profits, and many of the canal projects that subscribers had rushed to finance were foundering for want of funds. Canal meetings became very different from those of the end of the eighteenth century. When a meeting was held in 1811 to discuss, yet again, the proposed Bristol to Taunton canal, the speakers were full of pessimism and caution. They recalled the mania days, and gave endless accounts of projects that had failed. There was, for example, the Poole and Dorset Canal 'in which engineers had amused the Country with estimates to complete a canal sixty miles in length; but upon execution, it had turned out that *all* the money had been spent on making a few miles of a branch only of the Canal, which was now abandoned and of no use'. The meeting ended with a number of resolutions opposing the canal, 'which were unanimously carried'.[2]

So, within twenty years, canal finance had moved from a position when five times the authorized capital could be raised at one meeting to one when it could not be raised at all. Many canal promoters found themselves in the position of the proprietors of the London and Cambridge Junction Canal Company, though not all answered questions as to why the canal was not being built with such commendable brevity and honesty: 'I have only briefly to say, that it is solely because the money for carrying it into execution is not yet fully subscribed.'[3]

The 1790s may have been manic years, but sound schemes

[1] *Gentleman's Magazine*, 1792.
[2] *Taunton Courier*, 31 January 1811.
[3] *The Times*, 15 July 1813.

came up as well as unsound ones, and the money that went into the schemes was by no means all from the small-time gambler. It was true that the general lowering of share prices, from the early £200 right down to £1, had made it easier for the small gambler to join in the speculation, but even so, the major part of the finance came from investors who worked on a rather larger scale. Local interests were still represented, but increasingly the money came from the big merchants and merchant banks of the capital. One mercantile bank alone subscribed £23,000 to the Thames and Severn Canal scheme, and there were others putting in £10,000 each.[1] Not that speculators on this scale were necessarily any sounder judges of canal schemes than the man on the borrowed nag, but they seemed to be, and gave at least a spurious respectability to the proceedings.

Throughout the period, the overall picture of canal investment is one of steady change from the influential local interests, with their expensive individual holdings of the earliest promotions, through decreasing local involvement and increased speculative investment, finishing up with no investment at all. There were exceptions to the general picture. There were the purely industrial canals, built by the industries concerned with the industrialists' own money. There was also the small number of Government financed canals, paid for almost entirely from public funds. But the financial problems of the Canal Company were not over when the initial subscription was raised. They might have their authorized capital, but now they had to build a canal with it, and build it within the financial limits laid down by Parliament. Not many Companies managed to achieve this laudable end.

Having got the money, often with so little trouble, the Company needed someone to look after it, so that practically the first action of the new Company was to appoint a Treasurer, who, in effect, became the Company's banker. This was a position of enormous responsibility – hence the large sureties required, such as Wedgwood's £10,000. Since the Treasurer had such large capital sums to control, many Companies chose men who were already connected to banking institutions, and those who were

[1] *Felix Farley's Bristol Journal*, 21 June 1783.

not bankers when they started often became bankers in time. The Treasurers had all the normal banking functions to perform – handling and investing the cash funds of the Company, and paying the Company's bills as requested. Occasionally, like other small banks of the period, they mismanaged the funds and went bankrupt. This was a risk that the Company had to take, for even the regular banks were not completely reliable – there was still a firm distinction drawn, for example, between Bank of England notes and real money. However, the problem that arose most often to face the Company Treasurer was not his own shortage of funds but the Company's.

Most of the troubles that plagued the later canal projects were financial. Progress on some canals was held up because of engineering difficulties, but many more were delayed because of a chronic shortage of cash. The main cause of these financial difficulties was inflation. Canal companies worked on the system that has bedevilled many later, and supposedly more sophisticated, construction concerns – fixed-price contracts. They went to Parliament to receive their Act, which specified how much money they could raise for the building. This sum was based on the canal company's estimates of the cost, which in turn were based on current prices. Then, when the building actually started, prices would begin to rise – labour costs, costs of materials, land prices – everything went up. Telford gave some detailed figures for the Caledonian Canal, comparing costs of wages and materials in 1803 and in 1812. In 1803, labourers were paid 1s 6d per day, and those used to canal work 1s 8d to 2s; in 1812, the rate had risen to 3s to 3s 6d per day. Piece work for canal cutting was at 3d per cubic yard in 1803; in 1812 it had gone up to 4½d. Provisions went up in price: a boll of oatmeal that would have cost £1 in 1803, cost £1 16s in 1812. Other costs rose in similar proportions – costings worked out in 1803 had an unhealthy look by 1812.[1] The builders of the Caledonian Canal could draw on public money to meet the rising costs, the private companies had rather more difficulty.

The Huddersfield Canal Company was typical of many of the

[1] John Rickman (ed.), *Life of Thomas Telford* (1838).

canal enterprises that were formed in the 1790s, although it had a larger proportion of local investors than some of its contemporaries. It all began happily enough in 1794: the first general meeting was held, officers were appointed and the work was got under way. Three years later, the Company was starting to run into some financial problems. In theory, under the terms of the Act, they should have been able to raise more cash by a call on the original subscribers for extra payments. In practice, it was not so easy. They certainly tried, but in spite of cajoling, threatening and pleading, they found that too many of the subscribers preferred the forfeiture of their original investment to putting in any more cash. The Company struggled on. Then, in April 1797, the Treasurers sent a reminder that they were owed 'upwards of three thousand and four hundred pounds'.[1] The Company was overdrawn. Their first reaction was to order the sacking of the workmen, but the Company's engineer, Benjamin Outram, pointed out what the Committee appeared to have overlooked, that without workmen work would stop, expensive machinery would lie idle, and no one would ever get his money back. The Company then reversed their decision and looked for more sensible ways to get the money. They were able to raise a mortgage which kept them going along in spite of repeated scares of impending financial doom. In 1798, the Committee had to report that 'by the Bankruptcy of several of the Proprietors', the death of others and some emigration, they were unable to raise the money to pay their debts.[2] However, they still managed to stagger on until August 1799 when the canal workings were hit by floods. This time the Committee's report was gloomier than ever – due to 'the damage occasioned by the late unprecedented floods' they could no longer function. They could not afford to make the repairs, they could not repay their existing debts and they certainly had no prospect of finishing the canal.[3] In the circumstances, the only course open to them was to go back to Parliament to ask for authorization to raise more money. The

[1] *Huddersfield Canal Company Minute Book*, 7 April 1797.
[2] *Ibid.*, 16 February 1798.
[3] *Ibid.*, 23 August 1799.

money was duly raised, if rather slowly, and work was restarted. The canal was at last finished in 1811.

This story could be repeated, with minor variations, for many of the later canals. Not all were as lucky as the Huddersfield; some were only ever partly finished, others were just abandoned. More and more, the canal proprietors turned to their Treasurers for funds to help the company out of what they always assured themselves was a temporary financial difficulty. The Treasurer's banking role was confirmed and he was expected to stretch that role as far as providing an overdraft. The Committee of the Lancaster company wrote to their Treasurer and Banker, Thomas Norswick, to ask for an advance, and suggested, politely, that if he was either unable or unwilling to find the cash himself he should resign his post and hand over to someone who would loan the money.[1] Even the engineers occasionally tried their hand at raising money. 'I have been trying several of my monied friends . . . but the great gains in the Public funds is too tempting to make anything of them'.[2]

There were now, too, other and safer investments than canal companies. Rennie wrote again to Gregson in 1804: 'I am convinced that the necessary supplies cannot be raised unless a bonus of some sort or other can be given over and above £5 p. ct. for when nearly £5½ can be got in the public funds people will not lend it to canals for £5.'

The days when every canal scheme seemed to hold prospects of unlimited wealth died very quickly. The contrast between the different periods can be seen dramatically in two contrasting views of the Kennet and Avon Canal. In the first example, the Kennet and Avon's propagandist points out how easy the money will be to raise and just what a good investment it will be. He draws an analogy with what he considers a less attractive canal scheme, the Basingstoke Canal, pointing out that for even this inferior scheme £82,000 had been raised 'not by the neighbouring gentlemen only, but by persons totally unconnected with that

[1] *Lancaster Canal Letter Book*, 14 May 1797.

[2] Letter from John Rennie to Samuel Gregson of the Lancaster Canal Company, 7 October 1796.

8 The typical canal bridge.

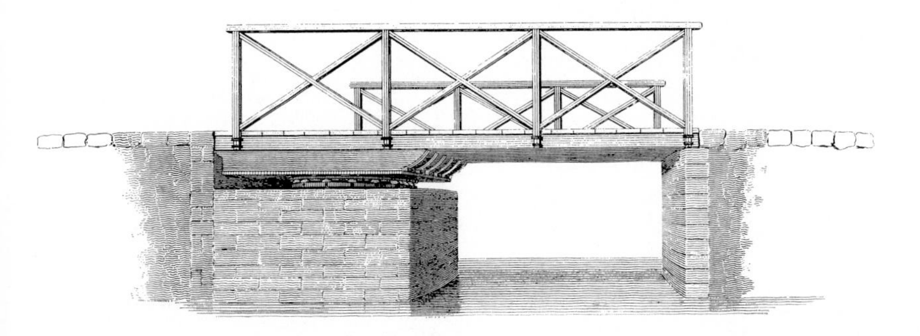

9 A Telford design for a swing bridge.

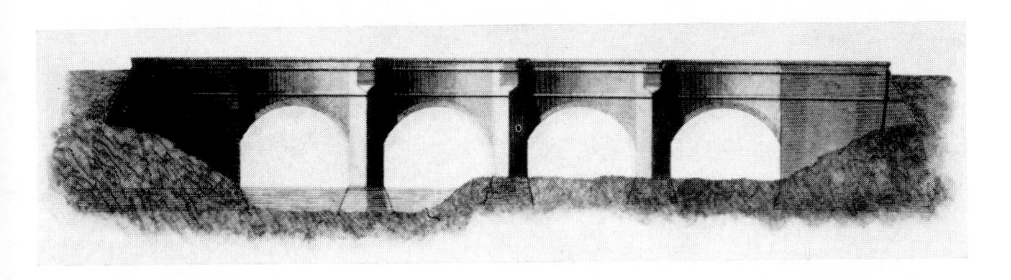

10 Elevation of the stone aqueduct at Kelvin on the Forth and Clyde Canal.

11 Three examples of simple inclined planes, using gravity or water for power.

country; merchants and bankers in London, and other great towns; men too well acquainted with the subject, and with the value of money, to engage in any undertaking which has not a reasonable prospect of advantage'.[1]

Just over twenty years later, the Kennet and Avon itself was being taken as an example, but was being used to discourage, rather than encourage, canal investment:

> Mr Sealy was of opinion that the Public had long been imposed upon, by engineers making plausible incorrect statements on the subjects of Canals–Mr Rennie, supposed to be one of the best, had estimated the Kennet and Avon at 400,000*l.* The money had been spent, and Mr Rennie was again called upon . . . the Company was obliged at last to go to Parliament for money to finish it, and it will cost a million of money.[2]

The speaker was unfair in placing all the blame on the poor engineer for not foreseeing the effects of rising costs. One should not feel too much scorn for Rennie's poor arithmetic. When one considers some of the more wildly inaccurate engineering estimates of the second half of the twentieth century, made by men who have the use of computers and a whole range of sophisticated techniques, Rennie may not seem to have done quite so badly.

The men who financed the canal projects did so for a variety of reasons and with a variety of results. It could be claimed that there was a certain natural justice at work – the financers of the earliest schemes often did so with the general good of the public as one of their aims, and had the double satisfaction of seeing the schemes completed and their own investment multiply in value. The financers of the later schemes, who went in as speculators, often lost heavily. This would, however, be a crude picture of the canal investor – there were speculators in the early schemes, and public-spirited men in many of the later. In between, there was a whole range of smaller investors who never made fortunes, but never lost them either. They might not do any better than they

[1] *Observations on a Scheme for Extending the Navigation of the Rivers Kennet and Avon,* 1788.

[2] *Taunton Courier,* 31 January 1811.

would have done had they invested their money in the public funds, but they had the satisfaction of being part owners of a great transport system.

There was one other important side-effect of the financing of canal construction. The men who handled the funds often gained improved expertise in the matter of dealing with the vast sums of money involved. And the sums involved were indeed large. During the mania years, from about 1788–95, over six million pounds was subscribed for canals, and another two million was borrowed.[1] Lessons could be drawn from the story of canal finance that would be applicable to later, and even larger, schemes. Some of this new financial expertise proved useful in the railway age that followed – many of the lessons were, unfortunately, ignored.

[1] T. S. Ashton, *An Economic History of England: The Eighteenth Century* (1955).

PART TWO

ENGINEERS AND ADMINISTRATORS

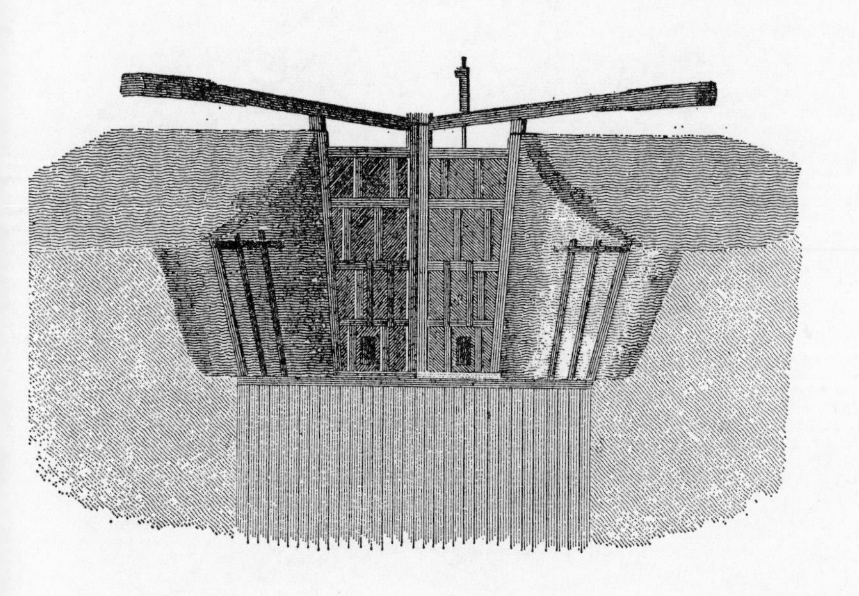

7

BUILDING A CANAL

The Act obtained and the money raised, the promoters and the politicians were able, at least temporarily, to take a rest and hand work over to the men who would be responsible for the actual building of the canal – the engineers, administrators, contractors and workmen. What sort of technical problems did these men have to face? What was there in the construction of canals that involved the employing of so many men for such a long time and at such enormous expense?

Given a level stretch of uninhabited land, then canal building is simplicity itself; all the canal builder needs is enough men with shovels and barrows to dig a trench across this open plain and then make it watertight. Unfortunately for the British canal engineer, he had to push his canal through well-populated areas and over a landscape that was rarely level. He also had to cope with other natural and man-made obstacles: he had to take his canal over rivers and streams, and he had to take roads and foot-paths over his canal.

The major obstacle facing the canal engineer was the change in the level of the ground. Faced with a hill, the engineer had three options: he could go round it, he could go up one side and down the other, or he could go straight through it. Most canals go up and down hills at some part of their course, and the most common device for overcoming the difference in level is the lock. Locks were in use as long ago as the fifteenth century, and they first came to Britain in the seven-teenth century. The principle of the lock is simple: it consists of a chamber, usually built of stone or brick, with watertight gates at either end and sluices that can be used to regulate the level of water inside the chamber. Although most have masonry chambers, there are examples, on the Shropshire Union Canal, of locks with sides formed out of cast-iron plates. Iron too was sometimes used instead of the more normal timber for the lock gates.

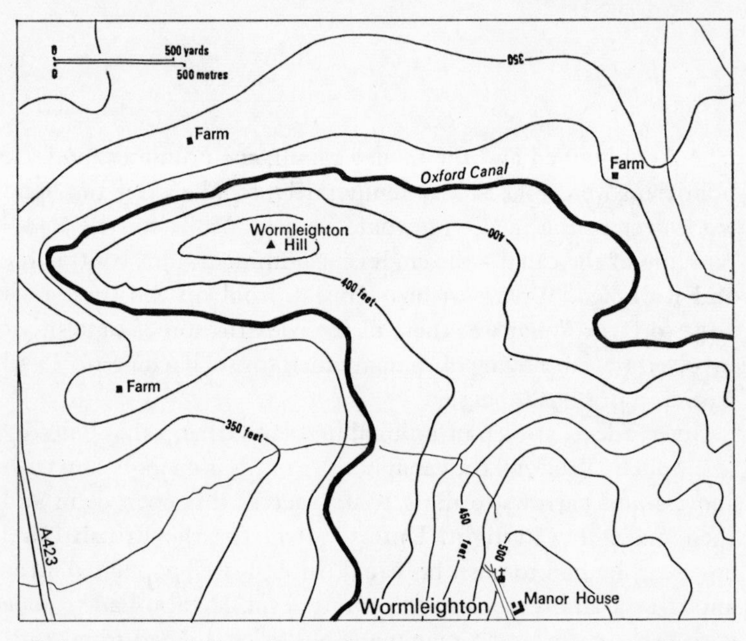

2 Going round a hill on the one level or 'contour cutting': the Oxford Canal at Wormleighton.

The making of a lock is a not particularly complicated operation, but even the introduction of such a simple device radically alters the type of workers needed by the canal engineers – instead of just needing diggers, he now requires masons, bricklayers, smiths and carpenters as well.

As the canal-building period progressed, so locks became bigger and more complicated. It was usual to limit the fall of water inside the lock to six to eight feet, but some canals were built with much greater falls – the biggest of all being on the Glamorganshire Canal which had a fall of fourteen feet six inches. Locks can be grouped together to form a 'flight', or they can be combined into a 'staircase'. A staircase consists of locks joined together, so that the top gates of one lock form the bottom gates of the next, and so on. The most impressive example in Britain is the Bingley Five Rise – a set of five connected locks on the Leeds and Liverpool Canal, which lifts the canal by fifty-nine feet two inches.

The lock was not the only device used for lifting boats up and down a hillside: there were also mechanical lifts. The most common of these machines was the inclined plane. This consists of a railed track up which the boats can be hauled. In the more sophisticated versions, the canal boat is floated into a large wheeled tub full of water, known as a 'caisson', and the whole is then pulled up by either water power or a steam engine. On some versions, there were two sets of tracks and two caissons – as the caisson with the boat went up one side, the second caisson came down the other and acted as a counterweight. There were not many inclined planes built. The first was constructed on the Ketley Canal in 1788, but there are no survivors today. Even fewer vertical lifts were built. The lift is simply what its name suggests: the same caisson principle is used, but instead of travelling on rails it is hoisted straight up into the air. It was not introduced until 1809 and the only working example of this particular device is the Anderton lift that connects the Weaver Navigation to the Trent and Mersey Canal, but this was not built until 1875. Mechanical lifts were very much the oddities in canal construction, and never became really popular with engineers.

The alternative to going round a hill or over it was to go through it by means of a tunnel or cutting. The techniques of tunnelling were borrowed from the mining industry and the final results, at least from the early period, seem now to be quite crude. These first tunnels were simply openings through the hills, and they were not supplied with towpaths: to pass through, the boatman had either to 'shaft' the boats or 'leg' them. Shafting was done by pushing a pole against the side of the tunnel and

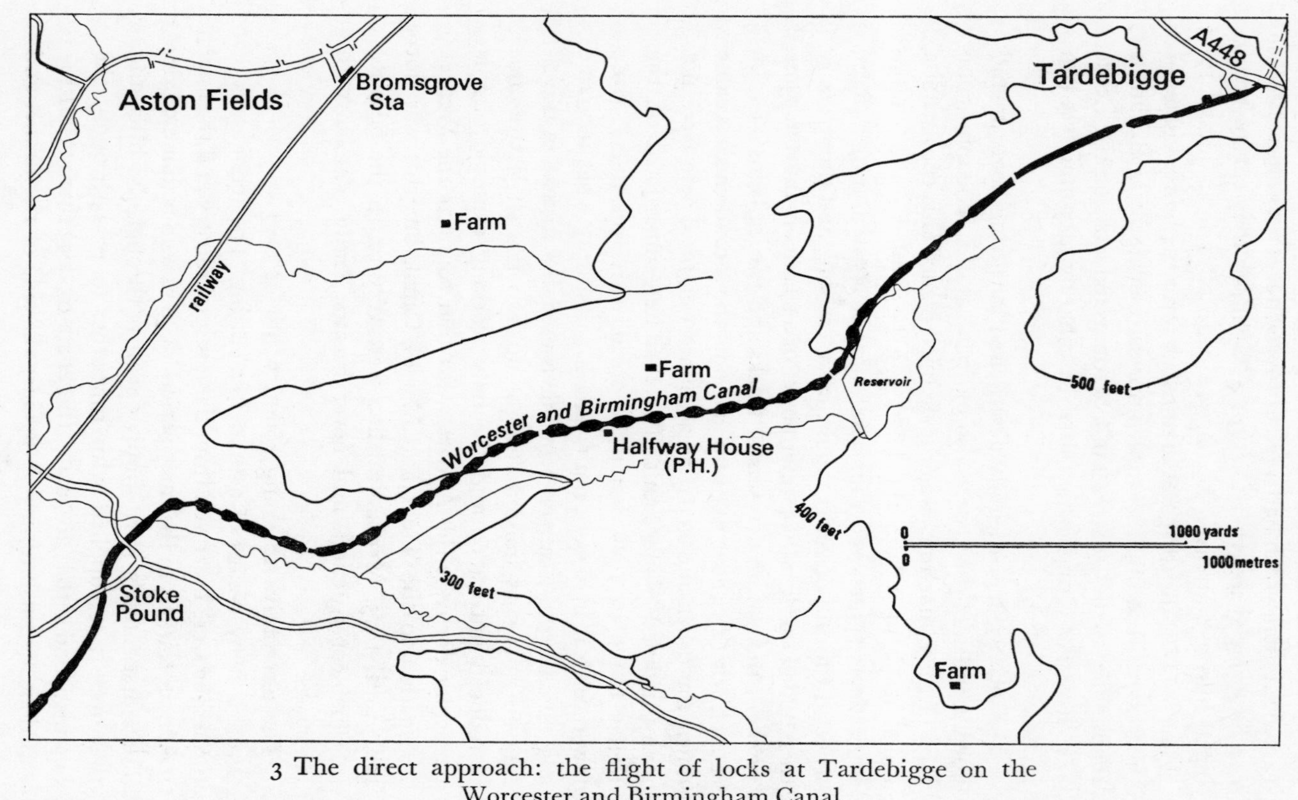

3 The direct approach: the flight of locks at Tardebigge on the Worcester and Birmingham Canal.

then walking the length of the boat, rather as in punting. Legging – or 'clogging' as it was known in the north of England – was an exhausting process in which the boatmen lay on their backs with their feet against the roof or the side of the tunnel and 'walked' the boat along. As the sides of the earliest tunnel were often formed by the rough rock, this must have been an exhausting job, and on the busiest canals it was taken on by professional leggers. Later tunnels were lined throughout with brickwork and provided with a towpath. The canal cutting is precisely what its name suggests – a gap cut through a hill.

Having got his canal over the obstacles of the hills, the engineer was still faced with the problems of the valleys. Rather than bring the canal right down to the lowest level, it might be more economical and a great saving in lock building to carry it across at a higher level by means of an embankment – an earth mound raised from the valley floor. Rivers could be crossed by aqueducts, which were usually built of stone, but occasionally of iron. The main difficulty facing the engineer when it came to building aqueducts was to ensure enough strength in the trough to contain the great pressure of water, which tended to force out the sides.

Because a canal passed through populated regions, it had to be liberally supplied with bridges. If, for example, the route cut across a farm, then the farmer would have to be provided with an accommodation bridge so that he could get from one part of his land to the next. These could be small stone bridges; or, in some cases, moveable bridges might be built. These would be made of wood and designed either to lift or swing to one side. More substantial brick or stone bridges would have to be built to carry roads across the canal.

The final, and perhaps the most difficult, problem facing the canal engineer was to ensure that enough water was always available for the canal. It might seem that in a notably damp climate like Britain's this would be the least of his problems. Not so: the passage of a boat through a canal uses up colossal quantities of water. Each boat going through the highest, or summit, level of a canal would use up two whole lock-fulls of waters. The lock-full which the boat uses to get on to the summit level eventually finds its way through the system and through the other locks and passes out at the end of the canal; the same thing happens to the lock-full used to get off the summit. In theory, this quantity of water should see the boat through all the rest of the system: in practice, some would always be lost through evaporation or leaks. Two lock-fulls of water might not seem a great deal, but when you consider that the average lock on the Grand Junction Canal holds 56,000 gallons, then you come up with the somewhat staggering figure of 112,000 gallons of water for each and every boat to pass through the canal, and that water has all to be replaced if the canal is not to run dry.

The first step which the engineer could take to ensure his water supply would be to make his summit level as long as possible, so that it could also act as a reservoir. This would, however, only be a start – it would not supply nearly enough water for the whole system. Natural springs and streams could be diverted as feeders for the canal, but these were usually more trouble than they were worth – their supply of water also carried earth and debris into the canal and caused silting. The main water supply had to come from reservoirs, which were specially constructed for the canal, together with an appropriate number of feeder canals. The Killington reservoir on the Lancaster Canal had a capacity of 700 million gallons. These, in turn, could be supplemented by pumping from wells or rivers.

As well as providing water, the engineer would be concerned to try to conserve the water he already had. Various devices were tried, such as side ponds at locks. These were constructed next to the locks. When a boat came through, not all the water was allowed to go out into the canal: part was diverted into the side pond, to be used partially to fill the lock when it was next needed. A variation on this was to provide locks in pairs, so that one lock could act as a side pond for its neighbour.

This is only the briefest of outlines of the sort of technical and construction problems that had to be faced by the canal engineer, but perhaps it is enough to show how far the actuality of canals differed from the theoretical idea of a long ditch full of water. The canal was a complex part of an intricate network, involving engineering work on a massive scale. The men who built this system could justly claim to be the foremost civil engineers of the age. The first of them was James Brindley.

8

JAMES BRINDLEY[1]

The greatest enthusiast in favour of artificial navigations that ever existed.

J. Phillips, *A General History of Inland Navigation* (1792).

If the Duke of Bridgewater was the 'father of canals', then James Brindley was the 'midwife', for he was certainly the man who brought the Duke's conception into the world. But somehow, the idea of Brindley as anything but a rough, down-to-earth engineer seems ludicrous, so it is probably as well to settle for calling him the first great canal engineer.

James Brindley was born in 1716 near the hamlet of Tunstead in the Peak District of Derbyshire. The family was poor, partly because the father was rather more interested in sport than in work, and was frequently lured away by the dubious delights of the bull ring about three miles from their home. As soon as he was old enough, the boy was set to work. He received no formal schooling of any sort, apart from a little rudimentary instruction from his mother. In 1733 he was bound to a seven-year apprenticeship to a wheelwright and millwright, Abraham Bennet of Sutton, near Macclesfield.

The millwrights' training was well suited to a career in

[1] The two main sources of biographical information on Brindley are Samuel Smiles, *Lives of the Engineers*, Vol. 1 (1874) and Phillips, *op. cit.* For a modern account of his part in the construction of the Bridgewater canal, see Hugh Malet, *The Canal Duke* (1961), though I have a different view of the relative importance of the roles of Brindley and Gilbert (see above, p. 86).

engineering, and many of the eighteenth-century engineers, including Rennie, began in this way. There was no acknowledged engineering profession, so the millwright, who had to be a self-sufficient workman in wood and metal, and had to develop a considerable practical experience with a variety of engines and mechanical devices, was as good a master as any. In theory, the apprentice learned his trade from the master – but in Brindley's case the master preferred drinking to teaching and the job of instruction was passed on to his journeymen, who were too often away working to bother about the boy. Consequently, Brindley found himself being given work which he had no idea how to perform, and which no one had the time or the patience to teach him. At first he made a mess of everything he tried, and was cursed by his master. But Brindley hung on, teaching himself by quietly watching and copying until he became a competent craftsman. Probably, this period of his life left a lasting mark on his character – he became a solitary worker, unwilling or unable to learn from others, preferring to solve problems on his own and in his own way.

His first major job as a millwright came when he was employed on a small silk-mill near Macclesfield. Very much to the surprise of his master, he completed the work perfectly, and at the celebration when the job was finished James Milner, the superintendent of the mill, showed his good judgement or clairvoyance by prophecying that Brindley would soon be a better workman than any of the others, which no doubt pleased Brindley, but could hardly have endeared him to his workmates.

Soon afterwards Brindley gave a convincing demonstration of his ability to take in details of quite complex machinery and carry those details around in his head without the benefit of any written notes or diagrams.

> His master having been employed to build an engine paper-mill, which was the first of the kind that had been attempted in those parts, went to see one of them at work, as a model to copy after. But notwithstanding this, when he had begun to build the mill, and prepare the wheels, the people of the

> neighbourhood were informed . . . that Mr Bennet was throwing his employer's money away, and could not complete what he had undertaken. Mr Brindley, hearing of this report, was resolved to see the mill intended to be copied: accordingly, without mentioning his intentions, he set out on a Saturday evening, after working the day, travelled fifty miles on foot, took a view of the mill, and returned back in time for his work on Monday morning, informed Mr Bennet wherein he had been deficient, and completed the engine to the entire satisfaction of the proprietors.[1]

Making allowances for the way in which stories tend to grow when related about famous men – a hundred-mile walk *and* a full engineering survey all in one week-end sounds a little excessive – it still shows that Brindley was a remarkable young man. His master, Bennet, obviously thought so, for soon afterwards he wisely retired to his bottle and left Brindley alone to run the whole shop. When Bennet died, Brindley was an experienced millwright, and set off for Leek in Staffordshire, where he began his own business in 1742. At that time, the local potters were just beginning to use flint in their manufacturing process, and Brindley found himself well and profitably occupied in building flint mills for the new 'flint potters' such as John and Thomas Wedgwood. He began to gain a reputation not only as an efficient workman, but also as an ingenious inventor, so that when the proprietors of the Clifton coal mines found they had difficulties in draining their pits they called in Brindley. He solved the problem of constructing a 600-yard underground channel, along which the water was drawn by a water-wheel powered by the River Irwell. It was, in a way, Brindley's first canal.

His reputation grew steadily and spread beyond the Staffordshire potters and their neighbours. In 1755 he was asked by Patterson of London to build the big wheels for their new silk mill at Congleton in Cheshire. When he arrived he found work had already started. The 'engineer' in charge of the mechanical construction took a very poor view of Brindley, whom he

[1] Phillips, *op. cit.*

described as a 'common mechanic', and refused even to let him see the plans. As the engineer in question had himself been unable to make any sense of the plans or to devise any way in which to make the machinery work, his high-handed attitude seemed merely absurd. Brindley complained, the engineer said either he went or Brindley went, and the engineer went. Brindley then took over, built the machines, which worked perfectly, and even suggested some improvements of his own. His reputation went up another notch.

About this time, Brindley became interested in the new atmospheric engines that were just coming into use. He went over to Wolverhampton, inspected a Newcomen engine, then came home and, as usual, decided he could do as well himself, if not better. In 1756 he received a commission to construct an engine for a Mr Broade. He began by trying, somewhat disastrously, to build an engine with wooden cylinders. But he continued, doing everything himself, until he started to get something more like a working engine. It took about a year to complete and was not an outstanding success. The trials began in November 1757 with eight days of 'Bad louk'; a little more tinkering by Brindley and he was able to record 'Midlin louk'; but it was not until March of the following year that he was able to write, triumphantly, 'Engon at woork 3 days', to be followed a little later by 'driv a-Heyd!' Although the engine gave him trouble, he was sufficiently pleased with the results to start work on a second, and then to apply for a Patent for his design.

In the midst of his work on steam engines, he was approached with the first proposal regarding canals. He was asked by Lord Gower to survey a possible line for a canal from the Trent to the Mersey. As far as one can see, Brindley had had no previous experience in surveying, but with his customary self-confidence he undertook the job, made his report, and went back to his engines. The decisive event in Brindley's life came in the Spring of 1759 when John Gilbert, the Duke of Bridgewater's agent, recommended consulting Brindley about the proposed new canal, partly, no doubt, because of Brindley's reputation in general engineering and mechanics and partly because he was one of the few men with any kind of experience in canal surveying. So Brindley duly set off to make an 'ochilor survey or a ricconitring'.

Exactly what part Brindley played in the planning of the Bridgewater Canal has now become a matter of controversy. Some historians, notably Hugh Malet, have argued that the main credit should go to John Gilbert – but the arguments in favour of this view seem inconclusive. It is difficult to see how Gilbert could have had time, as he was fully occupied in running the Duke's other affairs and coping with the constantly recurring financial crises. Certainly Brindley's contemporaries were quite clear where the main credit lay, although they were prepared to give others their due:

> It was the duke's great happiness to meet with a man of Mr Brindley's genius, which broke out like the sun from a dark cloud, he having been totally destitute of education: it was no less advantageous to the public, that under such a patron, Mr Brindley was called forth and encouraged, but other very ingenious men have assisted in carrying it on, particularly Mr Morris and Mr Gilbert.[1]

But it was Brindley's name that always came in for the most fulsome praise:

> An all-contriving power was given to the great Mr Brindley, sufficient to encounter all difficulties, and to remove the most perplexing obstacles. To his perforating hand the immense hills and stubborn rocks were no unsurmountable difficulty; and he could with the greatest ease carry water over waters.[2]

There are many examples that could be quoted, and it is difficult to believe that they were all unwittingly fooled. It is even more difficult to believe that they were fooled deliberately; even Brindley's detractors never cast any doubts on his personal integrity. Until more convincing proof to the contrary appears, it seems sensible to follow the traditional view and award the

[1] William Bray, *Sketch of a Tour into Derbyshire and Yorkshire* (1777).
[2] Rev. S. Shaw, *A Tour to the West of England in 1788* (1789).

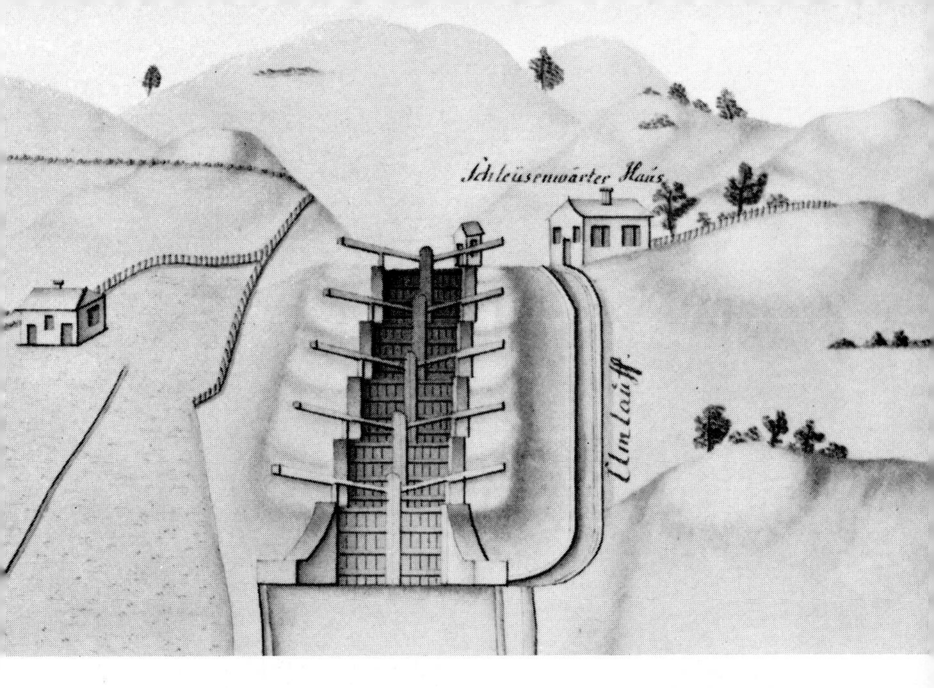

12 A rather curious view of the Bingley Five Rise on the Leeds and Liverpool Canal.

13 The two Harecastle tunnels: Telford's new tunnel is on the left, Brindley's original on the right.

14 John Smeaton.

15 James Brindley.

16 John Rennie.

17 William Jessop.

major share of the credit for the engineering of the Bridgewater Canal to Brindley – and certainly the canal itself follows one of Brindley's best known rules for canal building.

The first branch of the canal was level throughout the whole length from Worsley to Manchester, which accords with Brindley's predeliction for long stretches of dead water. He disliked rivers intensely, and described any water flowing downhill as a giant running along destroying everything in its path, whereas if the giant could only be got down flat on his back, all his destructive force would disappear. (Brindley's view of rivers was summed up in his famous reply, on being asked what rivers should be used for – 'to feed navigable canals'.) It has been argued that the design of the Barton aqueduct over the Irwell was outside his capabilities, as he had never attemped anything of the sort before. True enough, but then neither had anyone else, and lack of previous experience had never been a factor that weighed heavily on James Brindley.

Whatever Brindley's contribution to the planning of the canal, there is no doubt that he played a very large part in the organization of the work and the work force, and all observers were in agreement about the efficiency and impressive scale of the operations. The spectators were right to be impressed, for the work was superbly well handled, particularly since there were no precedents for what Brindley and his workers were attempting.

> Arriving at the head of the works, we were struck with the excellent and spirited appearance of active business; for the little village of Worsley looks like a river environ of London. Here is a very large timber-yard, well stowed with all sorts of wood and timbers for framed buildings, and building boats, barges, and all kinds of floating machines. The boat builder's yard joins, and several boats, barges, &c. are always on the stocks. Next to these is the stone mason's yard, where lie vast piles of stones, ready squared, for loading barges with, to convey to any part of the navigation where they may be wanted, either for building, or repairing of bridges, aqueducts, wharfs, warehouses, &c. The quarry is just by the mouth of the mine,

> and much is brought out of the mine itself, in working for the coals. Thus every part of the whole design acts in concert, and yields mutual assistance, which is the grand art of economical management.[1]

Arthur Young was describing the end of the construction period of the canal, but the same methods had been used throughout. As sections of the canal were finished, they were filled and used, so that materials from Worsley could be taken to the next set of workings. The necessary workshops, for carpenters and smiths, were housed in barges which were floated along the canal to keep pace with the workings. Other boats were especially adapted for taking waste from the diggings and for discharging it where it was likely to be needed, and, as Arthur Young noted with satisfaction, where particularly rich soil was brought up from the diggings, it was taken away to be laid on the Duke's lands.

The first great task that Brindley faced was the Barton aqueduct, and it says a great deal for the confidence of all three of the leading figures in the canal's construction that they were able to withstand the ridicule they received during the years of building. There is no record of what Brindley said or thought of his detractors, but it must have been particularly unnerving for a man with no education and no previous experience of this type of work to be told by men who had the best of educations and were supposedly learned that what he was attempting was impossible and that he was an idiot to try.

Apart from the aqueduct, Brindley also had to cope with constructing a waterway across the boggy ground that lay on the route to Runcorn – a stretch of land that was to give just as much trouble to a later engineer, Stephenson, when he came to build the Liverpool to Manchester Railway. An anonymous writer described Brindley's methods in a letter to the *St James's Chronicle* on 1 July 1765.

> He has finished the cut quite across Sale Moor, and will soon compleat it over the meadows on each side of the River Mersey;

[1] Arthur Young, *op. cit.*

> the entrance of which, from the low and boggy situation was, by men of common understanding, deemed his *ne plus ultra*. At this place, Mr Brindley caused trenches to be made, and placed deal balks in an erect position, backing and supporting them on the outside with other balks laid in rows, and screwed fast together; and on the front side he threw the earth and clay, in order to form his navigable canal. After thus finishing forty yards of his artificial river, he removed the balks, and placed them again where the canal was designed to advance.[1]

During his work on the Bridgewater Canal, Brindley had to solve every elementary problem of canal construction. His first contribution to canal engineering was to work out the best way of making a canal waterproof – the method known as 'puddling'. This technique involves kneading together clay, sand and water to make a semi-fluid 'puddle', which is then used, usually in two or three layers, to line the bottom and sides of the channel. No one ever found a better way of doing the job. Although the Barton aqueduct received all the attention when it was built, the high embankments that carried the canal to the Irwell were in their way just as impressive as engineering achievements. Brindley also devised a system of stop locks and drains whereby any one section of the canal could be isolated, so that repairs could be carried out.

Before the work on the Bridgewater was finished, Brindley found himself in demand from other canal promoters. Canals began to occupy more and more of his time, until eventually they virtually monopolized it. From being an ordinary, if more than ordinarily competent, millwright, he found himself elevated into the position of the country's leading expert in canal construction – for a time, in fact, its only expert in canal construction.

When the promoters of the Trent and Mersey scheme needed the services of an engineer, it was to Brindley that they turned. Although the first section of the Bridgewater had by then been successfully finished, Brindley was still very far from being the complete canal engineer. For a start, he still had not tried to build

[1] Quoted in Anon., *A History of Inland Navigations*.

a lock. When he did have to do so for the Runcorn extension, he showed that he still had a great deal to learn about canals. The episode of Brindley's first lock showed up the strengths and weaknesses of the man. Although the principles of lock building had been established for centuries, he was unwilling to take anything on trust – if locks had to be built on the canal then they would be Brindley locks built to a Brindley pattern. He made arrangements for a trial lock to be built. In theory, this was unnecessary – but Brindley had never seen a canal lock, and he was unable to read an account of one; so he would do what he had done before, follow his own instincts. It was by following this method that Brindley had, in the past, been at his most successful: this time he was not. It was only the Duke of Bridgewater's insistence that older methods be followed and adapted that saved Brindley from making a major engineering error by following his own plans for the locks. But at least, at the end of the affair, he had learned about locks in the only way he really understood – in practice.

Brindley's exalted position as the leading expert took him away from his accustomed surroundings and brought him to London to face the Parliamentary Committee (see Chapter 5). He did not enjoy the life of the capital, but he did his best to fit in with the strange surroundings. When he rode up to give evidence in January 1762, he was persuaded to invest in a set of new and fashionable clothes for the event. He was even prevailed upon to visit a theatre, and went off to see the great Garrick as Richard III. The evening was not a success: 'it had disturbed his ideas, and rendered him unfit for business. He declared, therefore, that he would not go to see another play on any account.'[1]

The Trent and Mersey plan presented Brindley with a new problem: his projected line involved tunnelling for almost three thousand yards through the hill at Harecastle. It was to be the country's first canal tunnel. There were the usual sceptics who said that such a tunnel was impossible, but, having proved the pessimists wrong in the past, Brindley found that this time he had supporters. People were beginning to express their faith in

[1] Phillips, *op. cit.*

James Brindley: he also had faith in himself, and the faith seemed to be justified as the work got under way, and the usual tourists came to look, wonder and report back to the London papers:

> Gentlemen come to view our eighth wonder of the world, the subterraneous navigation, which is cutting by the great Mr Brindley, who handles rocks as easily as you would plumb-pies, and makes the four elements subservient to his will . . . when he speaks, all ears listen, and every mind is filled with wonder, at the things he pronounces to be practicable. He has cut a mile through bogs, which he binds up, embanking them with stones which he gets out of the other parts of the navigation, besides about a quarter of a mile into the hill Yelden; on the side of which he has a pump, which is worked by water, and a stove, the fire of which sucks through a pipe the damps that would annoy the men, who are cutting towards the centre of the hill. The clay he cuts out serves for brick, to arch the subterraneous part. . . .[1]

But the work at the Harecastle tunnel proved rather more difficult than that optimistic account suggests: hard rock held up the tunnellers at some points; soft ground that became virtually a quicksand held them up at others. The water that collected in the workings was at first removed quite simply by pumps driven by wind- and water-mills, but, as the tunnellers drove deeper into the hillside, flooding became a serious problem and Brindley again had to get to work on an atmospheric engine. The primitive steam engine he constructed had to be kept working night and day to keep the tunnel drained. The four elements were not to be so easily controlled after all, and doubts about the progress were again raised. Josiah Wedgwood reported on a meeting of the Canal Committee:

> Mr Brindley was there & assured the Gentn. that he could complete the whole in five years from Xmas next, & there

[1] Anon., *A History of Inland Navigations*. Letter quoted, dated 8 September 1767.

> being a Gentn. present (not one of the Committee) who doubted of the possibility of its being completed in so short a time & seem'd inclin'd to lay a wager upon it, Mr B. told him, that it was a challenge he never refused upon anything which he seriously asserted & offer'd them to article in a Wager of £200 that he perform'd what he had said.[1]

Brindley lost the bet, but never had to pay – he died before the Harecastle tunnel was completed. By later standards, it was a crude enough affair – a rough hole hewn through the hillside, partly arched with brick but mostly bare rock. When Rennie went to survey it in 1820, he described it as 'being small and crooked with brickwork that was generally bad'. All of that was true, but at the time it was built it was an astonishing piece of engineering. It was the first tunnel of its type – and it was no minor affair. The main artery stretched for 2,897 yards through the hill, with tributaries branching off to the coal that lay under the hill, so that the mineral could be loaded straight into boats and out into the main channel. It was a great work – and it took eleven years to complete.

As the work on the Trent and Mersey and the Bridgewater continued, other canal schemes were begun, and all seemed to require Brindley's presence. The most important of the plans in which he became involved was the Grand Trunk – this was the name given to the canal system, which included the Trent and Mersey, and which was to link the industrial Midlands to the three great sea ports of Bristol, Liverpool, and Hull. The section joining the potteries to the Severn, the Staffordshire and Worcestershire Canal, presented Brindley the surveyor with a new and irritating obstacle. He needed to select a point to make a junction between the canal and the river, and he went first to the most obvious place, the thriving port of Bewdley on the Severn. This was the main centre for Worcestershire traders, and at its busiest periods as many as four hundred pack horses would be stabled in the town. The local inhabitants, however, could see no advantage in bringing in another waterway and suggested,

[1] Letter dated 12 March 1767.

somewhat impolitely, that Brindley should take his 'stinking ditch' elsewhere.[1]

The situation obviously called for diplomacy, but that was hardly Brindley's way. He set off along the river until he came to a spot where the River Stour emptied into the Severn, and there he chose to make his junction. It was an empty stretch of country, except for one inn, but Brindley decided that his canal would end in a major trading port that would rival Bewdley. He established offices and warehouses, docks and basins, and there the town of Stourport, Britain's first canal town, grew. Stourport and the 'stinking ditch' prospered – the trade of Bewdley declined. It is doubtful if Brindley ever felt much sympathy for its citizens.

Throughout the 1760s, Brindley was inundated with requests to work on canal projects as chief engineer or, at least, to give the promoters the benefit of his expert advice. He laid out the line for the Leeds and Liverpool Canal; he went to work on the Oxford and on the Coventry Canal. He was consulted on the Forth and Clyde Canal. He was even called in by the City of London to survey a possible canal from Sonning, on the Thames near Reading, to Monkey Island, near Richmond, but that scheme came to nothing:

> The bill met with such a violent opposition from the landowners, that it was defeated. Those fine gentlemen would not suffer their villas to be disturbed by noisy boatmen, or their extensive lawns to be cut through for the accommodation of trade and commerce; though it was from trade and commerce that most of their fine villas and extensive lawns had derived their origin.[2]

Brindley's existence became practically nomadic. He had very little time for any social life and indeed seemed to be more attached to his horse than he was to other people – his only big quarrel with John Gilbert was over the treatment of the mare. However, he did find time, when talking to John Henshall, the surveyor employed on the Grand Trunk, to notice that Henshall

[1] R. C. Gaut, *A History of Worcestershire Agriculture* (1939).
[2] Phillips, *op. cit.*

had a daughter, Anne. When they first met, in September 1762, Anne was still at school and Brindley got into the habit of filling his pockets with gingerbread when he went to visit Henshall. When she left school, Brindley proposed to her and was accepted. It was a bizarre courtship and an incongruous marriage between a rough fifty-year-old engineer and a young girl straight out of school. No one ever accused Brindley of having an excess of charm, and it is hard to avoid the conclusion that the father must have exerted some influence in favour of a match with such obvious advantages for his own career. However, the Brindleys settled down happily enough and the young bride was taken to Turnhurst, a large house with a suitably romantic view – it looked out on the work in progress on the Harecastle tunnel.

The demands on Brindley's time were excessive – everybody required his services on almost any terms, and he found it difficult to refuse any of them. The Companies were glad to get as much of his time as they could. The Coventry Canal Company, for example, appointed Brindley as Engineer and Surveyor at £150 a year, 'he undertaking to give at least two months attendance in the whole in every year'.[1] However, any canal company that signed Brindley up found that it did not always get quite what it expected: Brindley was a man who went his own way. The Oxford Committee began by taking a resolute stand with their wayward engineer. 'Resolved,' they wrote firmly, 'that Mr Brindley hath in no Degree complied with the Orders of the Committee.'[2] A year later they were taking a rather different line: 'Have read with great surprise and concern a Letter from you to acquaint us of your intended Resignation. We are very sorry that anything has happened that has given you Offence and shall always be ready to place the greatest Confidence in you. . . . Your letter mentions no Reason for this sudden step. . . .'[3] Brindley accepted the apology and withdrew his resignation – there is no record to show that he ever enlightened the mystified Committee as to why he resigned in the first place. Only one Company, the

[1] *Coventry Canal Minute Book,* 19 February 1768.
[2] *Oxford Canal Committee Book,* 6 September 1769.
[3] *Ibid.,* 12 September 1770.

Coventry, actually had the temerity to fire the great, but intensely irritating, Brindley.

The trouble was that Brindley was desperately short of time to give to all his different undertakings. Although an outwardly robust man, he suffered from diabetes, which remained undiagnosed until the time of his death. His constant travelling on horseback must have exhausted him, and not every job that he undertook received the care that it deserved. But it required a strong personality to stand up to Brindley, even when he was in the wrong. He met his match in John Smeaton, who with Brindley and a third engineer had been employed to survey the proposed Forth and Clyde Canal. Smeaton was more than ready to trade punches with Brindley:

> Mr Brindley recommends to begin at the point of partition, because, he says, it is his 'constant' practice to do so, and, in the present undertaking, it seems particularly advisable 'on many accounts': but pray, Mr Brindley, is there no way to do a thing right but the way you do? I wish you had been a little more explicit on the many accounts: I think you only mention one, and that is to give more time to examine the two ends: but pray, Mr Brindley, if you were in a hurry, and the weather happened to be bad, so that you could not satisfy yourself concerning them, are the works to be immediately stopped when you blow the whistle, till you can come again, and make a more mature examination? . . . That my brother Brindley should prefer the Printfield passage I can readily comprehend: a late author has very solidly demonstrated that every man, how great soever his genius, has a certain hobby horse that he likes to ride; a large aqueduct bridge over a large river does not happen to be mine.[1]

Brindley's endless journeying from canal work to canal work had its inevitable finale. In September 1772, he returned to his home at Turnhurst too ill to work. He died there on 27 September. There is a story, quoted by Smiles, that even when he lay

[1] Report of 28 October 1768 in *Reports of the Late John Smeaton, F.R.S.* (1812).

dying he was still besieged by canal promoters seeking his advice. One group told him, despairingly, that their canal would not hold water. 'Puddle it,' said Brindley. They replied that they had already done so. 'Then puddle it again – and again.' The story is probably untrue, but it at least demonstrates the incredible dependence which canal builders still felt on his word.

Throughout his life Brindley remained, in appearance at any rate, the simple country craftsman: 'he is as plain a looking man as one of the boors of the Peak, or one of his own carters'. His speech too remained heavily overlaid with the country dialect – you can almost hear the voice of Brindley from the few written notes, in which his spelling follows his pronunciation. But the simple countryman amassed a considerable fortune and rose to great importance. He was obstinate, single-minded and often narrow in his outlook – but this was, to some extent, an attitude that developed from the circumstances of his life. He was a man who had risen to pre-eminence by following his own ideas in the face of criticism – this led to a complete self-sufficiency but, because he could only work from his own experience, it led to inflexibility. His method of canal construction, for example, never really changed. He built, as far as possible, on the level, so that the canals he laid out tend to meander across the face of the countryside: today, the canal user finds the routes charming – the boatman of the eighteenth century who had spent the best part of a day going round in circles presumably had other thoughts.[1]

Brindley was a fascinating man in many ways. Because of his lack of a formal education, he had to devise his own method for solving engineering problems. He had an extraordinarily retentive visual memory, being able to look over quite complicated machinery and keep the details in his mind without the use of notes. Faced with a difficult problem, he would retire to his bed in a darkened room and lie there for as long as two or three days, emerging only occasionally to make a diagrammatic note as one part of the problem was solved. Then, when he was sure he had everything sorted out, he would leave the room and give a perfectly detailed account of his answer. Not that all the answers

[1] See, for example, the map on p. 76.

were entirely practicable – for a long time he harboured a pet scheme for building a floating canal to join England and Ireland – but the majority were.

At his death, most of the canal projects that he had begun were still unfinished, but he left behind the beginnings of the great canal network. If his canals now seem to be too narrow and too crooked, if the tunnel at Harecastle was crude and if even the famous Barton aqueduct was overshadowed by larger and more spectacular successors, it must be remembered that his were the first. He was the pioneer, and other engineers could learn from his examples. When he died, his friend Josiah Wedgwood wrote in a letter to Bentley:

> What the Public has lost can only be conceiv'd by those who best knew his Character & Talents – Talents for which this Age & Country are indebted for works that will be the most lasting monument to his Fame, & shew to future Ages how much good may be done by one simple Genius, when happily employd upon works beneficial to Mankind.
>
> Mr Brindley had an excellent constitution, but his mind, too ardently intent upon the execution of the works it had plann'd, wore down a body at the age of 55 which originally promis'd to have lasted a century.[1]

That can stand as a fair epitaph.

[1] 28 September 1772.

9

THE CHIEF ENGINEER

Civil-engineers are a self-created set of men, whose profession owes its origin, not to power or influence, but to the best of all protection, the encouragement of a great and powerful nation.

Preface to *Reports of the late John Smeaton, F.R.S.* (1812).

During Brindley's own lifetime, there were other engineers at work on other canals, but their activities tended to be overshadowed by the work of Brindley himself. His was the 'brand name' that all canal proprietors wanted on their product. After his death, a new generation of engineers came forward, the men who built the canals that spread across the country throughout the mania years of the 1790s and on into the early 1800s. Work on canals made some of them famous: William Jessop, who was responsible for completing one of the main arms of the 'cross', the main canal system that linked the four great rivers of England, Trent, Mersey, Severn and Thames, with the construction of the Grand Junction Canal; Benjamin Outram, who travelled throughout England and Wales building 'tramways' (the names 'tram' and 'Outram' have no connection beyond happy coincidence); John Rennie, father of an equally famous son, chief engineer for canals as far apart as the Lancaster and the Kennet and Avon. Other engineers worked on canal projects, but made their names in other fields. John Smeaton, for example, who first made his reputation when he built the Eddystone lighthouse. These were the men who became famous and rich. It would be impossible in the space available to give detailed biographical essays on all of them – it would also give an unbalanced view of canal building. True enough they were the ones whose names became linked with the more spectacular canal projects – we talk of Jessop's Grand Junction or Outram's Peak Forest – but to dwell exclusively on the chief engineers of these works would be grossly to undervalue the contribution of others, not least the men on the site who did most of the actual work. It would also be unjust to deal in detail with those few famous men at the expense of other equally competent, if less famous, engineers. So the object of this chapter is to give some idea of who the chief engineers were, but to concentrate more on the work they did and the kind of life they led.

Throughout most of the canal-building period, there was still no recognized form of training for a canal engineer, or indeed for any other kind of engineer; so the chief engineer would have arrived at his exalted position by any one of a number of different routes. John Rennie, for example, followed a career that was very similar in some ways to that of Brindley, beginning as a craftsman and ending as a professional engineer. He had the advantage, however, of having had rather more education than the almost illiterate Brindley. John Smeaton, on the other hand, came from a quite different background and followed a quite different career. He was born in Ansthorp, near Leeds, in 1724, and even as a child seems to have been fascinated by mechanical objects. His father was an attorney, who tried to get his son to follow him into the law – but the young Smeaton persuaded him that such a course would not suit the 'bent of his genius'. By the time he was eighteen he was a self-taught and proficient metalworker, who became interested in a whole range of scientific and technical subjects from instrument-making to astronomy. In 1753 he was elected to the Royal Society, and in 1759 received the Society's gold medal. His civil engineering career dates from 1755 when he was recommended by the Earl of Macclesfield, who was then President of the Royal Society, for the job of building a new Eddystone lighthouse after the old one had been burned down. From there he went on to a number of important civil engineering works: harbours, bridges and canals. For Smeaton, the work on canals was merely a part of a busy and varied professional career – a notable contrast to Brindley, who gave his canals all his time and enthusiasm. It is hardly surprising that the educated and urbane Smeaton had the self-confidence to stand up to Brindley, the acknowledged expert. The two men make a complete contrast in background and outlook: Brindley from a poor family, uneducated, devoted to canals; Smeaton from a wealthy family, with wide-ranging interests. They are the two extreme examples of canal engineer – most came somewhere in between.

Just as the men who became engineers came from different backgrounds, so their entry into the engineering profession and their training and preparation for the position of chief engineer differed. At Brindley's death, for example, a number of canal companies found themselves suddenly faced with the lack of a chief engineer, and men who had been surveyors or assistants suddenly found themselves promoted. As canal work became more common, men were able to get themselves taken on in junior posts, with the hope of learning the job as they went along and eventually finishing up in the chief engineer's chair. There was virtually no theoretical training, and what there was was not considered to be of any great value to the practical engineer. Sir John Rennie described his own training and the views of his father, the canal engineer:

> My father wisely determined that I should go through all the gradations, both practical and theoretical, which could not be done if I went to the University, as the practical parts, which he considered most important, must be abandoned; for he said, after a young man has been three or four years at the University of Oxford or Cambridge, he cannot, without much difficulty, turn himself to the practical part of civil engineering.[1]

A view of University education not entirely unheard today.

By whatever route a man arrived at the job of chief engineer, the job itself should, in theory, have followed a well-established

[1] Sir John Rennie, *Autobiography* (1875).

pattern. The chief engineer's responsibility was for laying out the main line of the canal, designing its main engineering features, preparing the necessary plans and specifications, and handing those over to the men on the spot who would then get on with the job of building the canal. The chief engineer would then make occasional descents on the works, as though from some Olympian heights, to see how the work progressed and, if necessary, to issue fresh commandments. In practice, the chief engineer found that things seldom, if ever, went that smoothly.

The first problem that the canal engineer had to face was how to deal with the Committee. The engineer was the qualified expert whilst the Committee was composed of unqualified laymen – but the Committee were the employers, and the engineer, however grand, only the employee. William Chapman was one engineer who fell foul of his Committee. He became involved in an argument over who should have the responsibility of hiring the Resident Engineer – the chief engineer's man-on-the-spot. Chapman argued that the decision was a matter of weighing up the applicants' technical abilities, a job which he alone was competent to undertake. The Committee argued quite simply that they were the employers, and they would hire and fire exactly as they saw fit. The argument grew so heated that Chapman tried to by-pass the Committee altogether, by appealing directly to the Company's shareholders for support. He wrote a pamphlet in which he stated his side of the argument, and quoted John Smeaton on the shortcomings of Committees in general: 'the greatest difficulty is to keep committees from doing either too little or too much – too little when any case of difficulty starts, and too much when there are none'.[1]

Chapman had no great success in his attempt to go over the Committee's head. Smeaton himself was just as forthright when it came to a Committee interfering in what he considered to be purely technical problems. He left them in no doubt as to how he viewed their different roles in the enterprise:

[1] William Chapman, *Address to the Subscribers of the Canal from Carlisle to Fisher's Cross* (1823).

If, instead of making plans, I am to be employed in answering papers and queries, it will be impossible for me to get on with the business. . . . All the favour I desire of the proprietors is, that if I am thought capable of the undertaking, I may go on with it coolly and quietly, and whenever that to them shall appear doubtful, that I may have my dismission.[1]

Smeaton had no worries in issuing a direct challenge, for he had many other sources of income. Others were not so fortunate. Most were aware that they were first in line for blame if things went wrong, and that they could be dismissed at any time. The wise engineer was careful to begin by establishing a good rapport between himself and the Committee. Usually he was successful.

The first part of the engineer's work – the laying down of the line for the canal – was the part in which the individual stamp of an engineer can most clearly be seen. Brindley and the other early engineers aimed at the flattest, which in practice meant the longest and most devious, line. In Brindley's case, this became a matter of principle – for most of his contemporaries it was a matter of necessity. The technology of canal construction was in too rudimentary a stage of development to allow for any other course; the engineers, not knowing a great deal about how to go through or over a hill or lacking the facilities to do so, went round it. Later engineers were able to take a more direct route: Samuel Smiles described the work of Rennie on the Rochdale Canal, built between 1794 and 1804:

In crossing the range at one place, a stupendous cutting, fifty feet deep, had to be blasted through hard rock. In other places, where it climbs along the face of the hill, it is overhung by precipices.

No more formidable difficulties, indeed, were encountered by George Stephenson, in constructing the railway passing by tunnel under the same range of hills, than were overcome by

[1] Smeaton to the Forth and Clyde Committee, October 1768, in *Reports of the late John Smeaton, F.R.S.* (1812).

Mr Rennie in carrying out the works of this great canal undertaking.[1]

The personal preferences and technical abilities of the engineer would, in a perfect world, be sufficient for the determination of the best line for a canal to take. Unfortunately for the engineer, the world was not that simple. Canals were not built across a deserted wasteland, and one of the first things that the engineer had to consider – or which, at any rate, the Committee would insist he had to consider – was the feelings of the more important of the local landowners. Sometimes the engineer would have to make a minor deviation: 'frequently, when the canal passed in sight of any gentleman's seat, [the engineers] have politely given it a breadth, or curvature, to improve the beauty of the prospect'.[2] In other cases, the changes necessary to keep the peace with a really influential landowner could involve rather more than a polite curve. Discussion about the best line for the Oxford Canal, which had originally been surveyed by Brindley, went on after his death. His successor, Samuel Simcock, set out the arguments for allowing the canal virtually to circle the hill at Wormleighton – the shorter, direct route would, he said, be more expensive and rather difficult to cut. Then he got down to what sounds suspiciously like the real reason behind the decision: the short route would 'be much more injurious to the Estate of Earl Spencer', whereas, when he came to consider the long route, he was able to point out: 'There is one pleasing circumstance in this part of the Navigation and that is, there will be no lock on the Canal thro' the whole Lordship, the navigators will have no business to stop . . . so that the apprehended Danger from the inroads of the Bargemen will be the less upon that account.'[3] The engineer was as aware as anyone else that the Company had no desire to quarrel with Lord Spencer. The long route was taken without further argument.

Even when it came to such matters as the design of an

[1] Samuel Smiles, *op. cit.*, Vol. 2 (1874).
[2] Thomas Pennant, *The Journey from Chester to London* (1782).
[3] *Oxford Canal Committee Book*, 7 February 1775.

aqueduct, which could have reasonably been supposed to lie entirely within the engineer's province, local politics, the need to appease special interests, cash problems, all had to be taken into consideration. Some concessions were harmless enough. Sir Francis Shipworth was given permission to paint the aqueduct at Stretton to fit in with the view from his house (alas, there is no record of the colour scheme he chose). John Rennie, who was responsible for Britain's finest masonry aqueducts, had his problems. When he was working for the Lancaster Canal Company, he found that the instructions he received about the bridges and aqueducts for the canal varied considerably. At times, the Committee were happy enough to indulge any whims and fancies, provided they were the whims of someone important: 'Mr Cartwright was mentioning that you had expressed a wish to have an ornamental battlement placed on the Hale Bridge . . . the Committee will be very ready to give any directions of this kind which you may think will give pleasure to Lord Balcarray.'[1] The Committee itself occasionally came in with a little advice on aesthetics: 'The flat cap on the top of the Pilaster at the Bulk Aqueduct will not have a good appearance. Could you not design something that will look more like a finished piece? The Work looks well and a little finish there would add to its beauty.'[2] And when it came to Rennie's masterpiece of canal engineering, the Lune aqueduct, the Committee were prepared to defend any expenditure. Samuel Gregson, the Company Secretary, wrote to Rennie:

> I have had several letters of enquiry from Mr Horner . . . he had heard that a great deal of money had been wasted on our Masonry. I endeavoured to explain these false Jokes away and recommended him to wait upon you to examine the design of the Lune Aqueduct. Then he would be able to judge how far the Committee ought to be censored for expending £127 – ! in ornamenting that great work.[3]

[1] *Lancaster Canal Company Letter Book.* Letter from Gregson, 26 July 1800.
[2] *Ibid.* Letter from Gregson, March 1794.
[3] *Ibid.*, 8 September 1800.

Six months later there was a drastic drop in the Company's funds and Rennie found that the instructions received by himself and Jessop, who had been called in to help in the crisis, were of a rather different tone. They were asked to resurvey the Ribble valley. The original plan had called for an embankment, another fine masonry aqueduct and locks; now the Committee wanted to consider a plain, unadorned aqueduct, possibly of iron instead of stone, and, above all, they wanted to know 'the best and *cheapest* mode of making a communication between Clayton Green & Preston, while an Aquedt of some sort or other is carrying on'.[1]

An engineer's life could be plagued by these social and financial pressures that reduced him to searching for compromises between the ideal construction and that which politics and money would allow. From that point of view, the engineers working on the earliest canals were more fortunate, for their works were less often hit by financial difficulties. But they had to work with unskilled labour and inadequate tools, which imposed their own limitations. One of the features of the later period was the large amount of technical literature that appeared, together with reports of various inventions, such as dredgers, for easing the labour of canal construction. But, in spite of all the compromises, the canal engineers were still, quite often, able to build what they wanted in the way that they wanted. All the great engineers had their specialities.

John Rennie's speciality was the aqueduct – the best of which, with their classical details, are among the eighteenth century's most impressive monuments. One side of his ability that has not perhaps received quite so much attention is his mechanical ingenuity – there is a splendid example of a typically elegant Rennie solution to a problem on the Kennet and Avon Canal. The difficulty was the common one of water shortage. He decided to build two pumping stations, to pump water into the canal – one at Crofton and the second at Claverton. The Crofton pumping house used a conventional pair of steam engines to lift the water, but at Claverton he had a different idea. The pumping house was

[1] *Ibid.* Letter from Gregson, 21 March 1801.

situated next to the River Avon, between the river and the canal higher up the hillside. Rennie's scheme was delightfully simple: a stream was led off the river and used to turn a water-wheel; this was geared to a beam engine which was used to pump water up the hillside and into the canal. So the power of water was used to lift water – the next best thing to perpetual motion. It is a tribute to the quality of Rennie's engineering and to the craftsmanship of his workers that the water-powered engine at Claverton was still working in the middle of the twentieth century.

William Jessop's speciality, if it could be called a speciality at all, was quietly, methodically and self-effacingly to organize a major canal project with the minimum of trouble to anyone else. He was many people's idea of the ideal chief engineer. He was a calm, modest and immensely sensible man – an attitude reflected in his work. His great project, the Grand Junction, is singularly free of any dramatic flourishes – its great tunnels at Braunston and Blisworth have no heavily ornamented entrances, no imposing façades – everything is as sensible and workmanlike as the engineer himself. Apart from being modest about his successes, he was always ready to admit to his mistakes – the most serious of which involved aqueducts, which were never Jessop's forte. Unfortunately this admirable, and, among engineers of that time, rare quality, misled many into underestimating his ability. Mr Robert Mylne, an irate shareholder in the Gloucester and Berkeley Canal Company, wrote to the Company's secretary in 1802:[1]

Well, my Good Sir,

I had your 2 Summonses to attend general meetings of 29 Sept. & 12 Octr. – Is there nothing going forward with the Canal? Is it to lay Dorment in a deserted State for ever? – It always was in want of a Chancellor of the Exchequer, & a better Chairman – Till this time, I have not had leisure to reflect on the miserable and misguided State of it, and its lamentable

[1] This is probably the same Robert Mylne who had been sacked from his job as Chief Engineer on the Gloucester and Berkeley, which would go a long way in explaining the ferociousness of the letter.

> fate – It is a work that requires other talent & knowledge than a common Canal Cutter – From the time of *Jessop*'s visit, I date its misfortunes. He and Dadford are mere drudges in that confined school; and both are without any sense of extended honour. The G. Junction is in a Dreadfull State, & required to have its difficult points all reconstructed. I have lately surveyed it. The Irish are advertising for a Resident Engineer. I think Dadford and they would fit one another to a T, for wrong heads and deficiency of knowledge.

Mr Mylne may have been master of a fine vitriolic rhetoric, but he was considerably less than just in his criticism of both Jessop and Dadford. Jessop, far from being a drudge, was a man who learned from his own mistakes and took a deal of trouble to listen to the advice of others. Rennie described a meeting with him:

> Since my last, Mr Jessop has been here and on conversing with him regarding the failure of the aqueduct over the River Darwin I find the lateral pressure of the water forced out the spandrill walls (which ultimately fell down) split the arch asunder longitudinally – this bridge has been repaired and they have adapted iron bars long and crossways same as I have drawn for the Lune . . . on showing him what I recommended in that way he said he so much approved of that mode he never would build an aqueduct of any magnitude without them.[1]

Jessop was certainly willing to learn; if he had a fault, it was in his overeagerness to don the hair shirt and take all the blame on to himself. Many of the troubles that beset him and all canal engineers came from the unavoidable gap between the instructions they issued and their execution, for they were seldom able to spend very much time overseeing the work themselves. Rennie stressed this point, whilst rather piously pointing out the superiority of his own methods over lesser engineers, particularly where they had had to skimp on materials:

[1] Letter to Gregson, 25 March 1795.

> I have hitherto made it a rule of my conduct to give such dimensions to works as to insure a certainty of success . . . many professional engineers have acted otherwise – hence the numerous failures that have lately happened in the various canal works &c throughout the kingdom . . . we have had no failure as yet, and trust while the plans are properly adhered to and the works well executed we shall have no occasion to be afraid.[1]

One engineer who became the complete specialist was Benjamin Outram. It became a common practice to build tramways, simple railways, between an industrial concern, such as a coal mine or iron works, and the nearest canal. When a tramway was being considered, the first name to come up was that of the tramway expert – Outram. The usual tramway was made of rails laid on stone sleepers, on which carts could be pulled by a team of horses. Occasionally, if the route lay up a steep hillside, as was often the case near the Welsh canals, the trucks would be hauled up by a stationary steam engine. It is particularly interesting to notice that the tramway system evolved as an early example of 'containerization'. The wagon was filled and drawn along the tramway to the wharf, where the container part of the wagon was lifted from the bogey and dropped straight into the waiting barge, where it fitted tightly in place. Later, during the years when so many canal companies were running out of funds, Outram found that his expertise was in even greater demand. It was very common for canals to be started from either end. When funds later ran out, the proprietors often found themselves with an embarrassing gap in the middle. One popular solution was to call in Outram to survey the canal and consider the feasibility of joining it together with a tramway. It is ironical that this solution, which seemed to offer salvation to hard-pressed canal proprietors, should prove to be the forerunner of their ruin – for it was on a Welsh tramway that Trevithick made his first successful experiments with a steam locomotive running on rails.

As well as planning and designing the canal, the chief engineer

[1] Letter to Gregson, 14 March 1795.

was expected to spend a certain amount of his time each year in inspecting the work in progress, reporting on what he found and attending general meetings of the company. This was all obviously necessary, but by the end of the century, when the mania was at its height, there was a huge increase in the numbers of canal projects – in the one year 1793 authorization was given for nineteen new canals to be built. It was easier to increase the numbers of canal projects than it was to increase the numbers of first-class engineers. Consequently, the men with big reputations found themselves constantly in demand – it was the Brindley story all over again, but multiplied. Each Canal Committee, naturally enough, wanted its full share of its engineer's time, and letters were sent continuously to keep him informed of progress, to ask advice and, with increasing frequency, to ask when he was coming to the works. As the Committee pursued the engineer with requests, the poor man himself was often trying to cope with a quite frighteningly hectic itinerary. When a letter from the Lancaster Committee finally caught up with him, Rennie wrote back on 13 November 1795 to explain why he had been such a long time in replying. He reported that he had just left Launceston in Cornwall, and had recently been at the Stourport and Birmingham canals, and had been to look at aqueducts on the Leominster canal. He was, he said, at present at York, but had to return immediately to London. He hoped to be up in Lancaster early in the new year, but he did have first to attend at two general meetings on 5 and 8 January. Bearing in mind that he was travelling on atrocious roads, in the worst weather, often visiting remote workings that could only be reached on horseback, that adds up to a gruelling period of work.

The successful engineer had to work hard, and seldom had more than a minimum staff to help with such work as surveying and drawing up plans. He was expected to involve himself in canal politics and canal finance. He even had to cope with the possibility of an early version of industrial espionage: a report was sent to the Oxford Canal Company, informing them of the activities of the engineers for the London and Birmingham Canal Company: 'Mr Walker and Mr Thomas have been in the Field for a fortnight. Would it not be proper to employ a spy on their operations that their parliamentary line may be known?'[1]

But if canal engineers were hard worked, they were also handsomely rewarded, gaining in terms of social status and hard cash. When Rennie debated the merits of a university education for his son, that in itself was a mark of the social progress that he had made. The engineer was able to move quite freely among, and associate on equal terms with, some of the best intellects of the age. He might not be equally at home in the fashionable and aristocratic world of London and Bath, but among the growing and increasingly influential groups of scientists, technologists and industrialists he more than held his own. When a portrait was painted of the most influential of these men, the canal engineers were fully represented.

The financial rewards for the successful engineer could be enormous. A man such as Jessop not only had his fees but was also a shareholder in various canal concerns and had shares in other industries as well. He had, for example, a considerable holding in a Derbyshire mining concern. Benjamin Outram's profits came as much from the Butterley Ironworks, in which he had a partnership with Jessop, as they did from tramway construction. John Rennie's account books covering the years 1794 to 1800 show an enormous increase in his income from approximately £7,000 per annum to £16,000. Out of this money he had, of course, to pay expenses, which were considerable, and salaries to his staff. Even so, the extent of his wealth can be judged by an entry for 1800, which showed that he paid out £11 for Christmas boxes. To earn £11 an ordinary canal labourer at that time would have had to work for six months. Brindley made a fortune but killed himself in getting it – later engineers were able to retire to live the lives of gentlemen.

But not all engineers were Rennies or Jessops. Some gave all their time to one project and faced retirement in rather different circumstances. The engineer during the early years of work on the Leeds and Liverpool Canal had been Mr Longbottom. He left the works when they temporarily closed during the depression

[1] Letter dated 19 July 1836.

years, and never returned to his old job. On 6 June 1800, he wrote to his former employers:

Gentlemen

Your Petitioner humbly presumes to lay before this General Meeting of the proprietors of the Leeds and Liverpool Canal such circumstances as he hopes will dispose them to take his case into consideration and to grant the object of his prayer.

Your Petitioner begs leave to remind this Company that he was for five years their Servant and Engineer. That during his Service he laboured with indefatigable Industry and exerted all his Abilities to promote the Interest of the Proprietors and he hopes he may say that when it is considered that in less than five years he compleated upwards of sixty Miles of the Leeds and Liverpool Canal for little more than one hundred and sixty thousand pounds his Abilities and Attention will not be called in question.

That since he was obliged to leave their service he has never been inattentive to their interest. He hopes he may state that the Proposals he has from time to time made to this Company of Particulars which he apprehended might be advantageous in compleating the work will fully justify this sentiment....

Your Petitioner begs leave further to state that in the decline of life, without Employment his Means of Subsistence are extremely slender; that tho' his Exertions thro' Age are not equally vigorous as in youth yet he trusts they may still be useful to this company and therefore he humbly prays

That this Company will have the Goodness to allow him a small annual stipend and he will very chearfully be at the Command of this Company in any Services to which they may think proper to call him.[1]

The Committee agreed to see if there was not something that could be done for Mr Longbottom. It was the following Spring before anyone got round to seeing the old man. He had died during the Winter. Historians tend to dwell on the successful and famous. There were others.

[1] *Leeds and Liverpool Canal Committee Book.*

10

THOMAS TELFORD[1]

Nor pass the tentie curious lad,
Who o'er the ingle hangs his head,
And begs of neighbours books to read;
For hence arise
Thy country's sons, who far are spread
Baith bold and wise.

Part of a poem by Thomas Telford, published in *Ruddiman's Edinburgh Magazine*, 1779.

Of the famous canal engineers it was Thomas Telford, who, apart from his canal work, closed the gap between the canal age of the eighteenth century and the age of nineteenth-century engineering achievement.

Telford's early life followed what, by now, must seem a familiar pattern – a poor childhood, training as a craftsman, introduction into engineering, and eventual fame and fortune. It is no wonder that Samuel Smiles became so fascinated by the lives of the engineers – they were all real, live evidence of the benefits of his famous doctrine of 'self help'.

Thomas Telford was born in Glendinning in Eskdale in the Lowlands of Scotland on 9 August 1757. Three months later, his shepherd father died and his mother had to move into a small

[1] The main sources for biographical information on Telford are his own autobiography, edited by John Rickman, *Life of Thomas Telford* (1838); Samuel Smiles, *op. cit.*, Vol. 3 (1874), and an excellent modern biography: L. T. C. Rolt, *Thomas Telford* (1958).

cottage, where they lived together in one room. The boy was fortunate that he had an uncle able and willing to pay for his schooling, although as soon as lessons were over what spare time he had was taken up in working for local farmers. The 'tentie curious lad' of Telford's own poem was himself.

Schooling over, Telford was apprenticed in Langholm as a stone mason – a craft at which he became highly skilled and in which he took a great deal of pride. Smiles tells a story of Telford paying a visit to his old home in Eskdale in 1795, when he was well established in his new career in engineering. There he met an old friend and fellow-workman, Frank Beattie, who had turned to innkeeping. ' "What have you made of your mell and chisels?" asked Telford. "Oh!" replied Beattie, "they are all dispersed – perhaps lost." "I have taken better care of mine," said Telford; "I have them all locked up in a room at Shrewsbury, as well as my old working clothes and leather apron: you know one can never tell what may happen." '[1]

Telford worked as a journeyman mason, and in 1782 set off for London, where he was employed on the building of Somerset House. He was a keen and ambitious young man who was already determined on a rather more promising career than that of stone mason – his aim was to become an architect. He studied whenever he had time and opportunity, and saved his money. He was fortunate in having a patron, William Pulteney, who was prepared to exert his influence to help Telford advance his career. To say he was fortunate is really an understatement: architecture, unlike engineering, was an established profession which was not easily entered. A young man needed patronage. When his work on Somerset House was over, Telford moved down to Portsmouth, where he acquired some of the practical experience and knowledge of harbour construction that was to be put to use later.

The big step forward for Telford came in 1786 when, largely through his patron's influence, he was appointed Surveyor for the County of Shropshire. This job brought him into an area in which the industrial revolution was already well advanced and

[1] Samuel Smiles, *op. cit.*, Vol. 3.

into contact with large-scale civil engineering projects – mostly road and bridge building. He rebuilt the county jail and even had the opportunity to indulge his taste for stylish architecture. Anyone seeing his church at Madeley needs no further evidence to decide that the most sensible choice Telford ever made was to abandon the career of architect for that of engineer. His new engineering career began in earnest when, in October 1793, he was appointed 'General Agent, Surveyor, Engineer, Architect and Overlooker of the Works' for the Ellesmere Canal Company. He continued to work, part-time, as County Surveyor, but canal work soon took over. He was doubly fortunate in his first canal job: he began work under the guidance of the best of the canal engineers and certainly the one most willing to give of his own time and experience, William Jessop, and he arrived at a time when the Company was ready to undertake the most ambitious civil engineering project of the century.

Telford's early work with the Ellesmere was that of a superior dogsbody, turning his hand to anything from the supervision of bridge building to debt collection. It was tiring work, but excellent training for a man in a hurry to learn. By February 1795 Telford was thought to have learned enough to be given his first job as chief engineer, and he took over work on the Shrewsbury Canal on the death of the previous engineer, Josiah Clowes. Almost immediately after joining the company, Telford was faced with the problem of deciding on a design for a new acqueduct to carry the canal across the River Tern at Longdon, the original having been washed away in a flood. The obvious choice for the former stone mason would have been a masonry aqueduct, but the leading proprietors in the Shrewsbury company were the great iron founders – William and Joseph Reynolds and John Wilkinson. The idea for an iron, instead of a stone, aqueduct was suggested by the Committee Chairman, Thomas Eyton, and was accepted. Telford took enthusiastically to the idea and went to work with William Reynolds on designing the structure. It was not a particularly impressive work, only just over sixty yards long and sixteen feet high, but it gave Telford the chance to work in a new material. He put the experience to good use.

Telford appears to have been an instant convert to the use of iron and at once began planning ways to use his new skill. That same February that he was begining to work on the Longdon aqueduct, he was asked to build a road bridge at Buildwas. By April he had submitted his design for an iron bridge. His next proposal for the use of iron was on a somewhat grander scale: it was for the aqueduct that was to carry the Ellesmere Canal over the River Dee at Pontcysyllte.

Just as there is some controversy over the allocation of credit for the Barton aqueduct, so there is argument about how the honours should be divided for Pontcysyllte. Charles Hadfield argues that a great deal of the credit should go to Jessop rather than Telford.[1] Contemporary commentators, on the other hand, had no doubts, and unanimously gave their acclaim to Telford. It seems unreasonable to try to show that they were either misguided or fools or both, unless there is very strong evidence to the contrary. What evidence there is does not seem to overthrow the accepted view. It is impossible to piece together the whole story, but what we know about the circumstances and the characters of the two main participants makes it possible to do a spot of reasoned guessing to fill in the gaps. What emerges looks at any rate a convincing if unverifiable picture of the events.

The original plan at Pontcysyllte called for a series of locks to bring the canal down the sides of the Dee valley, with the river being crossed by a three-arched masonry aqueduct. Stone had already been quarried in readiness by July 1795, when Jessop reported to the Committee that he had had a new plan presented to him that would replace the old set of locks and stone aqueduct by an embankment and an aqueduct in which the water would be held in an iron trough that would be carried on stone piers, 125 feet above the river. It does not require any great feat of imagination to decide who put the idea forward. Jessop never claimed the idea for himself and it came forward at a time when Telford, full of enthusiasm, was suggesting iron for everything. There is not much dispute here. The question at issue is, who was responsible for the detailed design and supervision of the

[1] Charles Hadfield, *Canals of the West Midlands* (1966).

construction once Jessop had got Telford's idea accepted? Most commentators give the credit to Telford and he, himself, admittedly writing many years after the event, claimed the credit. The suggestion has been made that by the time he came to write about the event, Telford was semi-senile and his memory was grossly unreliable. There is nothing in Telford's writing to suggest that this was the case. He reports firmly that the Committee was 'pleased to propose my undertaking the conduct of this extensive and complicated work'.[1]

Telford then goes into some detail about the methods which he employed during the construction, and again there is nothing to suggest faulty memory:

> The stone piers are 18 in number, beside the two abutment piers; they were all built to the level of 20 feet, and then the scaffolding and gangways were all raised to that level, and the materials being brought from the north bank, the workmen always commenced at the most distant or south abutment pier, receding pier by pier to the north bank; and by this ascending from time to time in their work, they felt no more apprehension of danger when on the highest, than at first on the lowest gangways.

He was proud of his safety record, if a little unfeeling about the solitary victim.

> . . . one man only fell during the whole of the operations in building the piers, and affixing the iron work upon their summit, and this took place from carelessness on his part.[2]

It is certain that Telford consulted with Jessop, who was, after all, the canal's chief engineer, and he mentions in his autobiography how grateful he was for Jessop's advice about the construction of the earthworks that brought the canal to the aqueduct. But, when it came to the aqueduct itself, it seems reasonable to suppose that Telford's own account is substantially accurate: he may not have been an experienced ironmaster, but

[1] Thomas Telford, *op. cit.*
[2] *Ibid.*

he was a brilliant and daring innovator. Pontcysyllte bears witness to an act of imaginative thought that seems more likely to belong to Telford than to Jessop.

It was the biggest undertaking of its kind of that period, and remains today the most impressive monument to the work of the canal builders. The opening ceremony on 26 November 1805 was a suitably grand affair.[1] The Shropshire Volunteers turned out for the occasion and brought along two brass field-pieces for firing salutes and their band for playing 'loyal airs'. Numerous dignitaries came to make speeches and to listen to others. The praises of Mr Telford, 'the projector of the aqueduct', were sung and there was much waving of banners, carrying messages varying from the poetic:

> Here conquer'd Nature owns Britannia's sway
> While Oceans' realms her matchless deeds display

to the more mundane, if rather more pertinent:

> Success to the iron trade of Great Britain, of which Pontcysyllte aqueduct is a specimen.

A procession of barges crossed the aqueduct and, after the speech-making, came back again: 'The discharge from the guns, as the procession returned, the plaudits of the spectators (calculated at full 8000), the martial music, the echo reverberating from the mountains, magnified the enchanting scene; and the countenance of every one present bespoke the satisfaction with which they contemplated this very useful and stupendous work.' After that, there was only a 'sumptuous dinner' for the dignitaries and roast sheep with 'an ample addition of beef and ale' for the workmen.

Before this event occurred, however, Telford had begun work on what, in terms of the size of the task, was the most ambitious of all British canal schemes – the Caledonian. This was part of a much wider project with which Telford was concerned – the plan for the development of the Scottish Highlands. In 1802 he carried

[1] The account of the opening ceremony from which the quotations are taken is that in the *Annual Register*, 1805.

out a survey for the Treasury Commissioners. The whole area was in an appallingly depressed condition – socially as well as economically. The scars of the disastrous uprising of 1745 were still felt in the land, and the evictions and appropriations of the new Highland landlords who had replaced the old-style chieftains left a population dispirited and desperately poor. Those who could afford to go left for the new colonies – the rest waited for something to happen. Telford proposed a massive work programme involving road and bridge building, harbour building and development, and the Caledonian Canal. The improved transport system would enable the region to grow in the long term, and in the short term, its construction would provide the work that was so badly needed.

The Caledonian Canal was planned to avoid the long sea route round the north coast of Scotland, by joining Fort William on the West Coast to Inverness on the East. Parliament authorized the canal's construction and guaranteed the necessary finance. It was to be canal building on a massive scale, and, if any canal can be said to rival the greatest of the railway works, then this is it. For a start, it was not a narrow barge canal, but a ship canal; then, it was to be built through the heart of the Scottish Highlands. Lastly, it was to be built by a work force that, initially at any rate, was totally untrained. Such a scheme, set in a wild and remote region of Britain, would have been inconceivable less than half a century before. When work on the Caledonian began, just over forty years had passed since the opening of the Bridgewater Canal. Canal technology had not just advanced – it had leaped into a new age. The Caledonian was possible because the means to make it possible were available: when Brindley began to work on the tunnel at Harecastle on the Trent and Mersey, he had to build a crude atmospheric engine to his own design; before Telford began any work on the Caledonian, he ordered three Boulton and Watt steam engines for draining – a 36 HP engine, one at 26 HP and a small 6 HP engine.[1] It is a fair indication of how far canal engineering had developed.

[1] *2nd Report of the Committee for Making and Maintaining the Caledonian Canal*, May 1805.

18 Thomas Telford.

19 Men of Science. Canal engineers are well represented in this group of eminent eighteenth-century scientists. Rennie stands in the foreground, while Telford looks up at him on his right; Jessop stands in the background. Other notables in the picture include: Boulton, Watt, Cartwright, Herschel, Cavendish, Jenner, Crompton, Trevithick, Dalton, Symington, Nasmyth and Davy.

20 The canal basin at Stourport, a bustle of activity on Brindley's 'stinking ditch'.

21 Telford's Caledonian Canal. This shows the basin at Inverness.

It was not any particular difficulties that make the Caledonian so remarkable, but the scale of the enterprise – the machinery that was brought in, the number of workmen involved, the sheer quantity of earth and rock that had to be shifted. It was so remote that it was hardly likely to attract the sort of tourist who went to have a look at other canals in the making. But Telford was a great friend of the poet Robert Southey, and together they went to see the construction when activity on the canal was at its height. Southey left a record of what he saw.[1] It is a doubly valuable record, for Southey was a professional writer who was able to describe the works with clarity and elegance, and, because he had the engineer with him, he could describe what he saw with accuracy. It is worth quoting Southey at length, for there is no comparable eye-witness account of just what was involved in a major canal construction. They began at Fort Augustus.

> Thursday Sept 16., 1819 – Went before breakfast to look at the Locks, five together, of which three are finished, the fourth about half-built, the fifth not quite excavated. Such an extent of masonry, upon such a scale, I have never beheld, each of these locks being 180 feet in length. It was a most impressive and rememberable scene. Men, horses, and machines at work; digging, walling, and puddling going on, men wheeling barrows, horses drawing stones along the railways. The great steam engine was at rest, having done its work. It threw out 160 hogs heads per minute [approximately 8000 gallons]; and two smaller engines (large ones they would have been considered anywhere else) were also needed while the excavation of the lower docks was going on; for they dug 24 feet below the surface of water in the river, and the water filtered thro' open gravel. The dredging machine was in action, revolving round and round, and bringing up at every turn matter which had never before been brought up to the air and light. Its chimney poured forth volumes of black smoke, which there was no annoyance in beholding, because there was room enough for it in this wide clear atmosphere. The iron for a pair of Lock-gates

[1] Robert Southey, *Journal of a Tour in Scotland in 1819* (1929).

was lying on the ground, having just arrived from Derbyshire. . . .

They continued on towards the eastern end of Loch Oich:

Some parts of the canal in which we walked this morning, were cut forty feet below the natural surface of the ground.

The Oich has, like the Ness, been turned out of its course to make way for the Canal. About two miles from Fort Augustus is Kytra Lock, built upon the only piece of rock which has been found in this part of the cutting – and that piece just long enough for its purpose, and no longer. Unless rock is found for the foundation of a lock, an inverted arch of masonry must be formed, at very great expence, which after all is less secure than the natural bottom. At this (the Eastern) end of Loch Oich a dredging machine is employed, and brings up 800 tons a day. Mr Hughes, who contracts for the digging and deepening, has made great improvements in this machine. We went on board, and saw the works; but I did not remain long below in a place where the temperature was higher than that of a hot house, and where machinery was moving up and down with tremendous force, some of it in boiling water. . . .

They then moved on to look at the work in progress between Lochs Oich and Lochy:

Here the excavations are what they call at 'deep cutting', this being the highest ground on the line, the Oich flowing to the East, the Lochy to the Western Sea. This part is performed under contract by Mr Wilson, a Cumberland man from Dalston, under the superintendance of Mr Easton, the resident engineer. And here also a Lock is building. The earth is removed by horses walking along the bench of the Canal, and drawing the laden cartlets up one inclined plane, while the emptied ones, which are connected with them by a chain passing over pullies, are let down another. This was going on in numberless places, and such a mess of earth had been thrown up on both sides along the whole line, that the men appeared in the proportion of emmets to an ant-hill, amid their own work. The hour of rest for men and horses is announced by blowing a horn; and so well have the horses learnt to measure time by their own exertions and sense of fatigue, that if the signal be delayed five minutes, they stop of their own accord, without it.

The scenes that Southey described, with tracks laid, excavators at work, and barrow runs, are strongly reminiscent of accounts of construction work on the early railways. From these scenes of intense activity, Telford and Southey went on to Fort William to view the flight of locks now known as 'Neptune's Staircase'. Southey's enthusiasm, like the lock, ran over.

We landed close to the Sea-lock; which was full, and the water running over; a sloop was lying in the fine basin above; and the canal was full as far as the Staircase, a name given to the eight successive locks. Six of these were full and overflowing; and when we drew near enough to see persons walking over the lock-gates, it had more the effect of a scene in a pantomime, than of anything in real life. The rise from lock to lock is eight feet, 64 therefore in all; the length of the locks, including the gates and abutments at both ends, 500 yards – the greatest piece of such masonry in the world, and the greatest work of its kind, beyond all comparison.

A panorama painted from this place would include the highest mountain in Great Britain, and its greatest work of art. That work is one of which the magnitude and importance become apparent when considered in relation to natural objects. The Pyramids would appear insignificant in such a situation, for in them we would perceive only a vain attempt to vie with greater things. But here we see the powers of nature brought to act upon a great scale, in subservience to the purposes of man: one river created, another (and that a huge mountain stream) shouldered out of its place, and art and order assuming a character of sublimity.

Whatever one thinks of Southey's views of the relative merit of the Egyptian Pyramids and Telford's canal, there is no mistaking the awe-struck tone of a sensible man faced with something far outside his everyday experience. The workings on the Caledonian were vast, and what is particularly striking is the degree to which canal work had become mechanized, with the huge dredgers floating along as each new section of the canal was flooded. The excavators and pumps were giants of their time, though they might seem quite insignificant today (a modern watcher might also be somewhat less sanguine about the harmlessness of all those 'volumes of black smoke'). But, with all the increased sophistication of techniques, digging the Caledonian called mostly for a lot of sweating men with shovels and barrows who could be called on to dig the deep, wide trench through the earth. The story of these men and their work will be told in more detail in the last section of this book.

The sort of technical difficulties that Telford had to face were exemplified by the construction of the sea-lock at Clachnacharry on the Beauly Firth at the eastern end of the canal. The shoreline at this point on the coast was very flat, so if vessels of any size were going to get into the canal then, as they could not get to it, the canal had to be taken to them. First of all an embankment was built out to sea. Clay was taken out of a near-by hill and carried out to the shore-line on a specially constructed iron railway. The bank was then extended outwards until it reached some four hundred yards from the high-water mark. Stones were laid on top and the whole left for six months to settle. The canal and the lock pit were then dug out of the mound. At first, the lock pit was kept drained by a chain-and-bucket system worked by a team of six horses, but, as the pit drew deeper, the horses had to be replaced by a 9 HP steam engine. The excavations were finished in June 1811 and the massive lock gates were hung.

The most impressive part of the whole undertaking was the section between Lochs Oich and Lochy, the deep cut which Southey described so vividly. But everything about the Caledonian was on a previously unconsidered scale. By the time of the opening in November 1822, the work on the canal had been continuing for almost twenty years and had cost over £900,000, of which nearly half was the cost of labour. But, although labour costs were the main item, it is worth noting that expenditure on machinery reached the very high figure of £121,000.[1]

It would be very pleasant to be able to say that, after all that expenditure of time, money and effort, the canal was a huge success. Unfortunately, it was not. The timber trade that was to have been its mainstay never materialized, and even if it had done so the navigation on the canal proved rather more difficult than had been expected. The passage through the Highland mountains all too frequently acted as a wind tunnel, against which boats could make no headway. But it was still a kind of triumph. It was certainly a personal triumph for Telford, which his friend Southey recorded for posterity, in a set of verses setting out Telford's engineering history:

> Telford who o'er the vale of Cambrian Dee
> Aloft in air at giddy height upborne
> Carried his Navigable road. . . .

and so on for rather a long time. These grandiloquent stanzas were, however, thought worthy enough to be inscribed on stone and set by the side of the Caledonian Canal. The waterway itself is a worthier memorial to the engineer.

During the time of the construction of the Caledonian, Telford worked on other canal projects, and quite a large part of that time was taken up with improving older parts of the system. It is a further mark of the success of the canals and the improvements in construction techniques that it took less than half a century for the earliest parts to prove inadequate to cope with the increased traffic. Telford worked on the Birmingham system and constructed a new Harecastle tunnel alongside Brindley's original. Where the first version had taken eleven years, the new tunnel, built next to the old, was finished in three and on completion was found to have been so accurately holed that a watcher at one end could see clear through to daylight at the other. And it was not only in Britain that Telford's engineering skills were in demand – he was also called in as engineer on the Gotha Canal, built between 1808 and 1810, in Sweden.

[1] *Twentieth Report of the Caledonian Committee*, May 1823.

Telford retained his interest in canals right up to his death in 1834, when the Railway Age was starting to come into its own, a development with which he was unable to come to terms. He was too old to change his ways, and could only see the railways as a threat to the canal system to which he had made such a massive contribution. But if, in this one instance, Telford showed himself to be lacking in vision, in every other respect he was the model for the new type of engineer – a man of great technical versatility, undeterred by lack of precedent for what he was attempting. His canal work can stand beside that of any other of the great canal engineers, but what gives him his unique importance is the great range of his activities. As important as his canal work was the work he did on road and bridge construction; and again he can stand comparison with the best. The Holyhead Road, with its two famous suspension bridges over the River Conway and the Menai Straits, is the work of a man clearly undeterred by natural obstacles – its conception has the qualities that the great Victorians, such as Brunel, brought to their work. Perhaps finest of all Telford's work was St Katherine's Docks in London. Telford was not noticeably successful as a traditional architect – as an industrial architect he is unsurpassed. But, in his own life-time, he received the most significant honour of all, when, in 1818, he became the first President of the Institution of Civil Engineers. It was more than a recognition of his own greatness, for it was a recognition of the importance of the new profession of civil engineering.

Telford, a man of paradoxes, is a difficult figure to characterize. He clearly had ambitions to be recognized as an artist, a man of sensibility. He worked hard at his poetry, with some success, but when he came to write his autobiography it turned out to be dull, pedestrian stuff, more of a technical treatise than the story of a man's life. He retained his love for architecture, but showed little talent for the profession. His Madeley church is built in the then fashionable classical style that would have fitted it, unremarkably, into a provincial city – in a small, country town it is merely an absurdity, lumpen and out of place. Yet, when he came to his industrial building, he found a mode of expression that had a simple elegance, a strength of form and structure, that perhaps spoke for the stonemason in Telford and which the, later, architect Telford would have buried under a mess of fussy and irrelevant detailing. This was perhaps the most important clue to Telford's life: his friendship with poets such as Southey was deep and real, but he could express himself most clearly in his own work when he reverted to the simpler styles of his youth.

Southey never indulged in any deep analyses of his friend, but was content to accept him as a happy, good-humoured companion: 'There is so much intelligence in his countenance, so much frankness, kindness and hilarity about him, flowing from the never-failing well-spring of a happy nature. . . . A man more heartily to be liked, more worthy to be esteemed and admired, I have never fallen in with.'[1] There was also a more serious side to Telford's character – his concern for 'improvement', coupled with a belief that men needed work and all men should be supplied with work to do. An engineer who writes poetry and designs churches seems an essentially eighteenth-century figure. An engineer pontificating on the moral virtue of work seems very Victorian. Telford seems to belong somewhere in between the two. In his autobiography he reprints a poem that he wrote 'in early youth, when the situation of the Author gave him little opportunity of being acquainted with English Poetry'. It seems typical of Telford – the subject matter is his old home of Eskdale, and it expresses the conviction that the growth of industry and an improved transport system bring with them not only prosperity but happiness as well. These are the thoughts of a Victorian engineer expressed in the language of a Georgian poet:

As o'er the land improving arts extend
Rejoicing ESKDALE feel their powers descend.
Stript of her cumbrous loads, her mountains rise,
While at their feet the peopled valley lies:
The less'ning woods, that dark and dismal frown's,
Now spread their shelter, not their gloom, around;
And where the boggy fen neglected lay,
Smiles the white cottage and the village gay.

[1] Southey, *op. cit.*

11

THE RESIDENT ENGINEER

A person capable of conducting the business of a Canal through, viz., that he is a good Engineer, can carry an Accurate Level, and has a perfect knowledge of Cutting, Banking, etc, and also that he is a compleat Mason.
Reference given for Archibald Millar, Resident Engineer on the Lancaster Canal, *Lancaster Canal Company Letter Book*, 27 January 1793.

The reference provided for this engineering paragon failed to mention three important qualifications for the job of resident engineer – diplomacy for dealing with irate or greedy landowners, authoritarian will for handling uncooperative contractors, and an indefatigable taste for travelling.

Once the chief engineer for a canal project had laid down his plans and drawn up his specifications he had finished the first part of his job. He could then hand the whole lot over to the Committee and leave for the next canal. The job of supervising the actual building of the waterway then fell to the resident engineer. He was at the top of a chain of command, in which everyone would have his place apportioned. In most companies, the resident engineer would be responsible for supervision of the building work and there would be a Secretary who would be responsible for administration. Both of these would have a staff of assistants. In the case of the engineer, the assistants would be responsible for supervising particular parts of the workings – one would be responsible for cutting, a second for masonry, a third for some particularly important structure, such as an aqueduct. The workers themselves would normally be hired by outside contractors, who would receive their instructions from the engineering staff. The secretary would have his own staff of clerks and accountants. The success or failure of a canal-building project would depend, to a large extent, on the ability of the resident engineer and the efficiency with which the organization operated. John Rennie tried to ensure the smooth running of the Kennet and Avon Canal construction by dividing the work into two sections, and appointing a resident engineer to each part. He then laid down an exact set of rules and job specifications. He began by pointing out one important difficulty to be overcome: 'the difficulty of obtaining Resident Engineers and Agents of abilities, Integrity & Experience is more than what is generally imagined – indeed from the numerous Canal Works now in execution throughout the Kingdom, it can scarcely be otherwise expected'.[1] He then went on to give his detailed proposals:

1. To divide the works into 2 sections. Clerks should be chosen who are not just capable of keeping books; the ideal should also be capable of 'purchasing the land and agreeing for Damages, Settling the Situation for accommodation bridges &c &c – in this way I think his time may be fully employed'.
2. Each division should have a resident engineer. Each resident engineer should have an agent and assistant for superintending masonry and another for earthworks.
3. The resident engineer's job would be to superintend the works and to suggest improvements wherever possible. He should have an office near the centre of the line. The agents should look after particular jobs and also keep a regular check on the numbers of men, horses and carts in use on the site.
4. All agents should keep a record of their own work.
5. The resident engineer should be responsible for purchasing timber, ironwork and so on.

[1] *Kennet & Avon Canal, Western Sub Committee Minute Book.* Letter from Rennie dated 22 June 1794.

6. The principal engineer should provide designs and specifications, give instructions on the way the work is to be carried out and inspect the works before each quarterly meeting of the Company.

That was the theory, and it all sounded very reasonable. In practice the demarcation between jobs was never that distinct – the engineer, instead of only supervising and instructing the contractor, became involved in administration, whilst the secretary found himself involved with everything and everybody, and often the contractor finished up taking precious little notice of either of them.

The resident engineer was usually a man who grew up with the canal age. The first were usually employed on canals because of their abilities as surveyors or masons. Having got the necessary experience, they then moved around from one canal project to the next, though not with the mobility of the chief engineer, who would be able to hold on to many canal jobs simultaneously. For the resident engineer, there was only one job at a time and that usually lasted for many years. During the mania years, a good engineer was very much in demand. Later, as the age come near to its end, work was more difficult to come by. Samuel Hodgkinson had handbills printed setting out his previous experience – thirteen years as resident engineer on the Kennet and Avon, six years on the Birmingham and five years surveying for various companies: the Worcester and Birmingham, the Stratford upon Avon, the Dudley and the Stourbridge. In September 1820 he sent one of his handbills to the Committee for the Gloucester and Berkeley asking if there was a job available. He added some further information by hand, pointing out that he had spent twenty-five years as 'Canal Engineer in Surveying and executing new Canals and repairing old ones and it will appear on inquiry that the works executed under my direction have been done in a Superior way, are of considerable magnitude and have given general satisfaction to my employers'. The employers' note scribbled on the back, was terse enough: 'to be answer'd saying we are Supplied'. The life of a Resident Engineer could become very hard.

An alternative to canvassing around the canal projects was to advertise in the press. Mr Edson placed the following advertisement, and must surely have found some customer to take advantage of his astonishing array of talents:

> CANAL NAVIGATIONS, &c: D. Edson *Engineer* and *Architect*, having been many years engaged on *Canal Navigations* in the *North of England* has taken up his residence in *York-Street, Bristol*, and solicits the patronage of the Nobility, Gentry and Public in general, in surveying, estimating and executing Canal Navigations, Roads, Bridges, Sea-Docks, Sea-Banking, and making Pleasure-Grounds, with the several Appendages necessary thereto; by whom every Branch of Engineering and Surveying is estimated and executed on eligible terms.[1]

The necessarily long stay at each job meant that the resident engineer had very little opportunity to get any further in his career. The canal building age itself was so short that by the time his first major job was completed, he would find very little prospect of any more work in canal construction. Consequently the resident engineer became neither rich nor famous, and most of them are now little more than names in the pages of a minute book. A few were sufficiently odd to excite some comment. One of these was Matthew Davison. He started his working life, with Telford, as a mason in Langholm. He stayed with Telford, working as supervisor on the Pontcysyllte aqueduct, after which he suffered what he considered to be the great misfortune of a return to his native Scotland when in 1805 he was appointed resident engineer on the Inverness end of the Caledonian Canal. Southey heard a lot about the curious Mr Davison when he went to Inverness:

> The Canal here was under the superintendence of Mr Davison, a strange, cynical, humourist, who died lately. He was a

[1] *Felix Farley's Bristol Journal*, 1 December 1792.

Lowlander who had lived long enough in England to acquire a taste for its comforts, and a great contempt for the people among whom he was stationed here; which was not a little increased by the sense of his own superiority in knowledge and talents. Both in person and manners he is said to have very much resembled Dr Johnson; and he was so fond of books, and so well read in them, that he was called the Walking Library. He used to say, of Inverness, that if justice were done to the inhabitants there would be nobody left there in the course of twenty years but the Provost and the Hangman. Seeing an artist one day making a sketch in the mountains, he said it was the first time he knew what the hills were good for. And when some one was complaining of the weather in the Highlands, he looked sarcastically round, and observed that the rain would not hurt the heather crop.[1]

Davison was not typical of canal engineers, although there was also another aspect of Davison that Southey did not mention: he was exceptionally good at his job. Telford gave a major part of the credit for the construction of the sea-lock at Clachnacharry to him. Another resident engineer achieved fame, but not for his canal work. He made his name because he found time in a busy life to take more than just a passing interest in his surroundings. William Smith was trained as a surveyor before he was taken on to canal work. He spent six years from 1793 to 1799 as engineer on the Somerset Coal Canal, and later worked for a while on the Kennet and Avon. As he went round the works his interest in canals lessened as his interest in the rock strata that were laid bare by the canal cutters grew. By the time he left canal work he had begun to work out the geological significance of what he had seen, and set out a theoretical basis for his observations. It was this work that earned William 'Stratum' Smith the title of 'father of English geology'. William Smith, however, could hardly be taken as a pattern for the way of life of the canal engineer any more than Matthew Davison could. Precisely because they never achieved the fame of the

[1] Robert Southey, *op. cit.*

chief engineers, the resident engineers are largely unknown as individuals, though their work was hardly less important than that of their famous superiors. To get a picture of the working life of the resident engineer it is necessary to piece together what we know from a variety of different canal records.

As with the chief engineer so with the resident engineer, the first task to be faced was getting to terms with the Committee who employed him, though this was perhaps even more important for the resident engineer – he had the Committee permanently at his elbow and had no other canal project to get him away from the more troublesome moments. Smeaton explained the engineer's difficulty perfectly:

> To fit a man fully for this employment, requires so great a number of qualifications, that I look upon it as impracticable to find them united in one person. I therefore take it for granted, that he will, of course, be materially deficient in something; and as such, there is the greatest difficulty in the world to preserve good understanding between the resident Engineer and the Committee who directs him. His post is the post of envy. Not only all the inferior departments are ambitious to be practical engineers, but even members of the Committee have a propensity that way too; by which means all become masters, and he who ought to be so, being deprived of authority, it is easy to figure the confusion that may follow.[1]

Having reached some sort of agreement with his Committee, which might or might not turn out to be workable, the engineer was ready to begin the job itself. The first thing that had to be done before any section of canal could be cut, was to decide what land was needed and then acquire it for the Company. The survey of the land was the engineer's responsibility, and was a very important part of his work. Although the Parliamentary line had already been laid down, it gave only a rough indication of the route. The final choice lay with the resident engineer, who had to find the best and the cheapest line, taking into account both the cost of the land and the cost of cutting. The actual

[1] Quoted in William Chapman, *op. cit.*

purchase was sometimes undertaken by the engineer, sometimes by the Secretary.

The ideal was to reach a speedy agreement with the landowner concerned and, obviously, to get the land on the best possible terms. The Parliamentary Act would have laid down the mechanism for settling disputes, but this involved lawyers and they cost time and money. So it was very much in the interest of the Company to settle with the minimum of fuss, and unless they wanted to see the price of the land pushed up, they would not want to appear too eager. Archibald Millar, engineer on the Lancaster Canal, reported to his Committee that he had set out a section of canal, and urged them quickly to get on with acquiring the necessary land:

> When the Land occupiers or Land Owners understand that you must have their land & immediately for to let your Contractors get on, then at that time every one of them will try to impose, and they are certain not to settle but at the most Extravagant Rates.[1]

Quite a few landowners did try to take advantage of finding themselves in a seller's market. Some of the smaller ones found that they had been too greedy, and the canal company simply ignored them and went round the overpriced patch. Millar wrote scornfully of one such place: 'I think the man who possesses said Orchard (which does not deserve the name of an Orchard) a besotted fool.'[2] Sometimes, however, the surprised engineer found a landowner quite willing to settle, but on rather bizarre terms – like the vicar who went away happy when the engineer agreed to provide him with a new graveyard by way of compensation. In most cases, the Company had few problems with the small landowner – it was the large ones who were the potential trouble-makers. The engineer's role then reverted to his proper one of giving disinterested advice on the best line to follow and leaving the Committee to negotiate terms. Dealing with the country landowners, unless they had something

[1] *Mr Millar's Reports to the Lancaster Canal Committee,* 1 October 1793.
[2] *Ibid.,* 12 November 1793.

specific to gain from having a canal through their land, sometimes required more tact than even the most careful secretary was either able or willing to expend. These country squires and aristocrats could hardly believe that it was possible for their land to be bought without their consent. Lord Spencer's Agent replied to the Oxford Committee's request for a meeting to discuss land purchase, and one can almost hear the howl of outrage at this proposed insertion of crude commerce into an established area of ancient privilege:

> I am favoured with your letter of the 25th February but till I have communicated the subject to Lord Spencer . . . I am not qualified to give you any decisive answer. Nor, be it as it may, can I possibly attend immediately or indeed for some time, and I should think it would not have been amiss if the Company had offered me a little previous notice, as it can hardly be expected to be convenient for me to attend at a moment's warning. It would seem by what you say that so very considerable a quantity of land as 16 or 17 acres is the object of your attention and so far as I can judge from the complexion of your letter it is not in Lord Spencer's power to refuse the Accommodation however inconvenient or disagreeable it may be to his Lordship to furnish it. I shall be obliged to you to inform me, as I really do not know, if this is the case, and if Lord Spencer has no other privileges or Command over his own property than that of treating for the Price of any such parts of it, much or little, as the Company may happen to discover are convenient for their use.[1]

And this was only the Agent's reaction! Sometimes these negotiations became so difficult and protracted that the engineer was left in the unenviable position of being unable to proceed, while his work force stood idle – a position that the Trent and Mersey engineer had to face:

> Ten miles only remain to finish this great work, and to open the long-desired communications between the ports of

[1] *Oxford Canal 'Oddments Book'*. Letter dated 1 March 1787.

> Liverpool, Bristol, Hull, and the interior parts of this kingdom, which the company are proceeding to accomplish with all possible expedition. But it is much to be regretted that this communication, after so much labour and expense, will still be incomplete, if the obstruction given to it by Sir Richard Brooke, at Norton, in Cheshire, is not removed.[1]

Surveying the area through which the canal was supposed to pass, setting out the line of the canal, and conducting part, at least, of the negotiation for the purchase of the land took up a good deal of the engineer's time, and also involved him in much travelling around the countryside. Once the land was acquired, however, the workers could move in – and that presented the engineer with a whole new set of problems.

The first of these came because the workers themselves were not normally under the engineer's direct control, but worked for the contractors. A report of June 1796 showed that at that time there were thirty-five different contractors at work on the Lancaster Canal – and that was by no means an unusually high number. The largest of these, Stevens, had 150 men at work; the smallest, Pat O'Neil, controlled a total work force of two diggers. Some contractors were responsible for specific jobs – bridge building, cutting and puddling a certain length of canal – while others took all the work for one section, doing everything themselves. The engineer had the job of seeing that each of them, big and small, did his part of the work correctly and promptly, and that all the different parts were coordinated. The ideal was seldom, if ever, achieved. When the technical difficulties that were bound to arise in any canal work were added to the difficulties of working with such disparate forces, then the situation could exasperate even the most patient of engineers. When complaints began to appear of the slow progress of tunnelling work on the Leeds and Liverpool, the chief engineer, Robert Whitworth, wrote to the Committee to explain how the two sets of circumstances were combining against him.[1] First,

[1] *Williamson's Liverpool Advertiser*, 12 October 1775.

[2] *Reports of Robert Whitworth to Leeds and Liverpool Canal Committee*, 1 June 1792.

22 Telford's spectacular masterpiece, the aqueduct at Pontcysllte. The picture also shows the earthworks that brought the canal to this point.

23–24 Rare photographs of an Outram tramway in use. It carried coal from Denby, and horse-drawn trains ran on it until 1908. The top picture shows the iron rails on their stone sleepers; the lower shows the technique of 'containerization' in use.

there were the technical reasons – the cutting was through bad ground, 'nothing but mudd and running sand', and they were constantly having trouble with land slips. Then there were the contractors; on top of the usual sets of complaints that applied to most of them, two in particular had caused a lot of trouble: 'John Parkin, a conceited, Ignorant old man that would not be guided, and John Layburn who run away.'

All engineers had problems with contractors, and they all stemmed from the central problem of lack of direct contact with the workmen. All orders had to be given through the contractors but, when the large contractors were involved, the engineer sometimes found himself completely frustrated because he could not find anyone to whom he could give the orders. Not surprisingly, he sometimes gave up looking for the contractor's agent and gave the orders direct – and then waited for the outcry. If he was lucky, the Committee would back him up, as the Lancaster Committee did when the situation arose between their engineer, Millar, and the contractors Pinkerton and Murray:

> You complain our Agents gave directions to your Workmen instead of yourselves or your Agents. You well know your attentions have been very little employed in respect to the Masonry, and your only Agent (Mr Tate) if he knew anything about the business had little time to attend to it, there were none but Workmen to direct, and our Agents were in fact obliged to Act as Superintendants for you.[1]

Archibald Millar and the Lancaster Committee had a running battle with many of their contractors, but Pinkerton and Murray were very much at the top of their black list.[2] A large proportion of the correspondence recorded for the Canal was between various agents of the Company and the contractors, and the same theme dominated Millar's reports. It is hardly surprising that a constantly heard cry was for the contractors to be taken off the works and for the control to be given directly to the engineer. This was impracticable for a number of reasons:

[1] *Lancaster Canal Letter Book*. Letter dated 4 January 1794.
[2] See Chapter 14.

workers, for example, preferred to stay with contractors who could move them on from canal to canal, and it was simpler to deal financially with a few contractors than with a few hundred men. Failing that remedy, the engineer would press for what he considered the next best thing, the replacement of large contractors by small concerns: 'Great Contractors without they have Agents equal in Authority and as many of them with a proportionate number of Men so as to equal a given number of small Contracts – never should be acceded to by the Committee for great undertakings like the Lancaster Canal.'[1] Millar, who made that report, spelled out in the bluntest terms what he saw as the advantages of using small contractors: 'No fine Law contracts are necessary on such occasions. If the man does not behave as he ought to do, drum him off. These are as safe contracts as any.'[2] This is very much an engineer's-eye view of things – the Committee continued to prefer the administrative simplicity of a few contracts to spending their time settling affairs with a multitude of small units.

The principal part of the engineer's work was supervision, where the main difficulty lay with the enormous area he had to cover, with workers possibly spread over many miles of the canal. There were also many different activities to command his attention. One day he might have to sort out problems with the carters – the men who brought the essential materials to the workings, stone from the quarries for bridging, timber for lock gates and so on. He would perhaps find that they had been using narrow-wheeled vehicles, which were speedier and more manœuverable than the broad-wheeled wagons that were specified, but which tore into the country roads and converted them into an impassable quagmire. When he had finished dealing with them, he would then have to spend some time pacifying local residents who were not overjoyed at the destruction of their roads. Complaints about damage caused by canal works were an inevitable part of the construction period, and a standard procedure for paying off complaints was in force. A

[1] *Mr Millar's Reports to the Lancaster Canal Committee*, 29 October 1793.
[2] *Ibid.*, 26 August 1794.

typical directive from the Committee instructed the engineer 'to wait on Mr Radcliffe and to explain to him . . . what has been usual for other Canal Companies to pay for similar damages and to endeavour to convince Mr Radcliffe that no Damage hath been done that could be avoided.'[1]

Another day might be spent in measuring the work – as most contractors were paid by work done rather than by time spent this was an important item, and one that frequently led to disputes. Somehow the engineer's measurements never came out quite as high as the contractor's. This, unless it could be resolved, led to yet more complications, with independent assessors having to be called in to arbitrate. Sometimes the argument became still more serious and led to charges against the engineer. During work on the Caledonian, the engineer, Easton, was accused of 'fraud and collusion with the Master Workmen in measuring the Cubic Contents of the Spade-work'. The case was heard, but no witnesses that were called corroborated the charge and Easton was cleared.[2]

In between dealing with problems of men and administration, the engineer even found some time to deal with engineering problems. One area where the engineer frequently had direct control was in the use and maintenance of pumps and engines. A contract for cutting the Gloucester and Berkeley Canal, for example, specifies that 'wheresoever it shall be necessary to lift any water, in order to keep the works dry, in cutting the said Canal, the said Canal Company shall provide proper Pumps, Engines &c and to Work the same so that there shall be no hindrance to the Workmen in digging and removing the Earth.'[3] During the canal-building period the steam engine was still in a very early stage of development and responsibility for its use was not an altogether unmixed blessing. The engines needed careful attention: 'The crank of the fly wheel shaft broke yesterday, which drove the Diggers out of the Bottom of the Pit. Having a spare crank by them, the engine in a few hours was got to work

[1] *Huddersfield Canal Company Minute Book*, 31 March 1796.
[2] *Tenth Report of the Caledonian Canal Committee*, 20 May 1813.
[2] Contract with William Montague, dated 7 October 1794.

again.'[1] But when the engine was working, the engine-man might not be:

> I hope we shall be able to begin the Piling about next Monday, if our Steam Engine People keep Sober – Mr Richards the New Man has kept the can to his head for 3 or 4 days continually. He promises never to be drunk again. I gave him a very good lecture yesterday and plainly told him the next time should be the last with him in this Company's service.[2]

Although there were Superintendents for the different sections of canal construction, ultimate responsibility for the quality of the work rested with the resident engineer, so everything had to be inspected – masonry work had to be checked, and the all-important puddle had to be carefully tested to make sure it was waterproof. Weather played an important part in the engineer's life, particularly during the early stages of construction. During this period all materials for the canal had to be brought in by road, and a long spell of rainy weather reduced everything to a muddy chaos – work and inspection stopped when the engineer found it literally impossible to reach the site. Later, when part of the canal was filled, men and materials could be moved by boat and rain became less of a worry.

On major works on the canal – such as a big aqueduct or long tunnel – the engineer would normally have the help of an assistant who would remain on the site to supervise the work. In the case of the big aqueducts on the Lancaster this was particularly necessary, as the Company was responsible for the first stage of construction: the contractors only came in when the foundations had been raised up above the high water mark. The life of the assistant was not particularly comfortable. Millar gave instructions for building what was to be the assistant's home, while the Lune aqueduct was being built. He ordered the assistant, Exley, 'to erect a shade, proposing it to cover a Saw Pitt, a Carpenters Shop, a room & kitchen for himself & a Store room. The kitchen will serve the Steam Engine man also. These

[1] *Lancaster Canal Letter Book*. Letter dated 26 March 1794.
[2] *Mr Millar's Reports* . . ., 25 March 1794.

buildings I have desired to be enclosed with the Slabs, that will come off the Foundation frame & Piling: And the posts to be made of the inferior Balks that may be unfit to be used under the Masonry. I have desired that the room for Exley to sleep in may be floored, and that it, with the Kitchen and Store Room be separated with Slabs, and that a fire place be made in Exley's room & in the kitchen also.' He concluded, rather chillingly, 'I have designed them to be raised on the most frugal Plan.'[1]

The assistant was prepared to accept life under these rigorous conditions, because the practical experience could raise him in the engineering hierarchy. Exley, for example, who was 'said to understand masonry & was bred a carpenter' had begun his canal career as an assistant supervisor of masonry, and must have seen the chance to act as supervisor for such an important work as the Lune aqueduct as a major step forward. Even so, his life was extremely rigorous. The workings at the Lune were subjected to frequent flooding, and on one occasion not only the workings but the living quarters as well were swamped. There was little peace, for work went on night and day. There must have been times when Exley, with little company besides the drunken engine man, wished he had chosen a different occupation.

The life of a resident engineer was an exhausting round of inspection, coercion, and conciliation. He expected to get the blame if things went badly wrong at the works, but when the canal was finished he could expect very little of the credit. He was, however, comparatively well paid. The pay rates on the Lancaster Canal show why there was such keenness to climb the engineering ladder, and it is also notable just how far payments had come since the rates were settled for the Trent and Mersey in 1766.[2] Millar received a salary of £400 a year, while his principal assistant got £250. Lower down, there is the superintendent of masonry, who worked for two guineas a week, and Thomas Fletcher was induced to come across from the Leicestershire Canal to superintend cutting and banking for a salary of £100

[1] *Ibid.*, 24 January 1794.
[2] See p. 35.

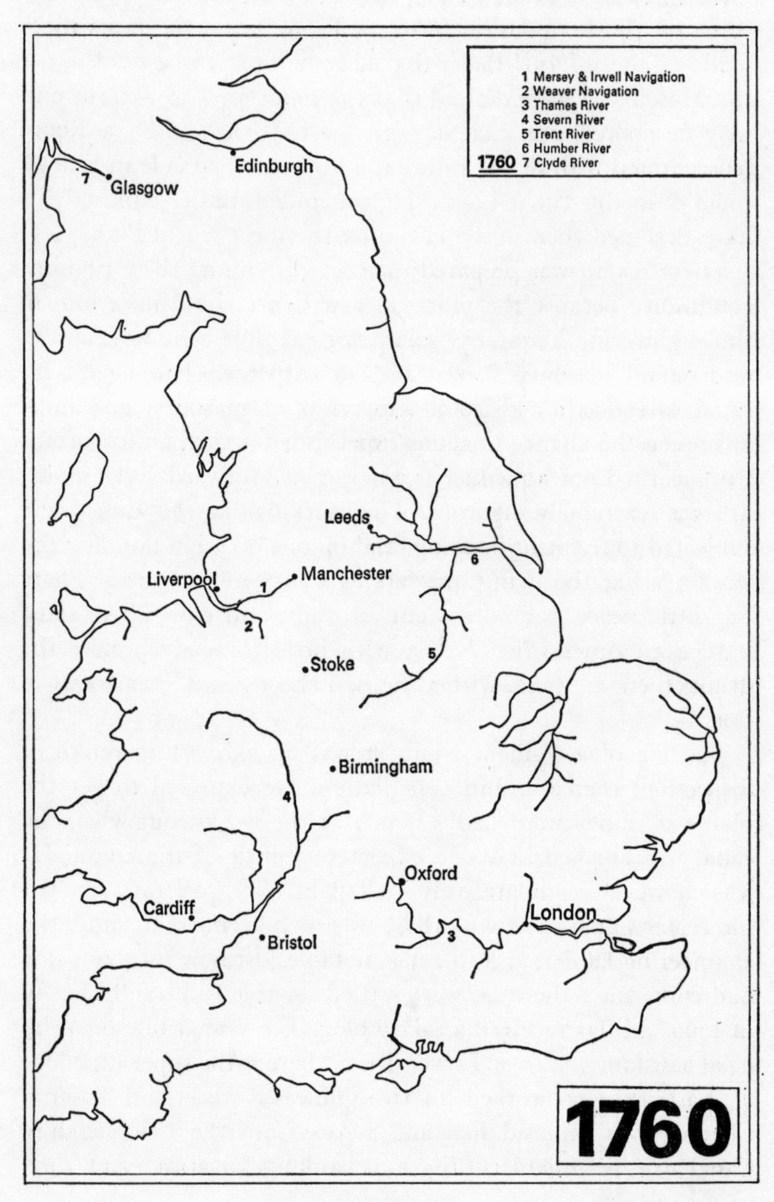
1 Mersey & Irwell Navigation
2 Weaver Navigation
3 Thames River
4 Severn River
5 Trent River
6 Humber River
1760 7 Clyde River
Edinburgh
Glasgow
Leeds
Liverpool
Manchester
Stoke
Birmingham
Oxford
Cardiff
Bristol
London
1760

4 The waterways in 1760.

a year. If these salaries, set in 1791, are compared with the very top position then they may not seem to be very high. The chief engineer received two guineas a day plus expenses for his initial survey and was then paid a salary of £600 a year for five months' work in Lancashire. Bearing in mind that the chief engineer could, and did, take on other work for the remaining seven months, he was clearly very much better off than the resident engineer. But, seen from the other end of the scale, Millar's salary looked more than reasonable. To a workman earning about 12s a week, Millar's £600 would have been an undreamed of fortune. The resident engineer may have been overworked, but he was hardly underpaid. The rates on the Lancaster were reasonably typical for the time, though in some instances a resident engineer could be called in on a day-to-day basis. Thomas Dadford was taken on by the Brecnock and Abergavenny Company to oversee work on the Clydach aqueduct, which he had designed, and for this he was paid £2.12*s*.6*d*. a day.[1]

It took many years for a man to rise to the position of resident engineer, and very few ever got beyond that point, but at least they always had the prospect in front of them that one day they might be appointed as chief engineer. The administrators who shared many of the engineer's problems had only hard work, and no prospects – however remote – of a golden future.

[1] *Brecnock and Abergavenny Canal Navigation Committee Minute Book*, 8 December 1794.

12

ADMINISTRATION

This has been a day of complaints and I am heartily tired of them.

Samuel Gregson, Secretary to the Lancaster Canal Company, in a letter to Archibald Millar, 9 April 1796.

Samuel Gregson was to the administration of the Lancaster Canal what Archibald Millar was to the engineering, and, on reading through his correspondence, one gets the impression that every day was a 'day of complaints'. He was the canal company's professional worrier. During the period of canal construction, anyone who felt he had any sort of grievance against the company or against one of the company's employees went, as a matter of course, to the representative of the canal Committee – the gentleman usually known as the Secretary or as the Chief Clerk or Clerk Accountant. To simplify matters we shall call him just 'the secretary'.

In theory, the organization and administration of the canal was the responsibility of the elected Committee; in practice, the day-to-day handling of affairs was left in the hands of the secretary and his staff. The Committee took the major decisions, but relied on the expert advice of their engineers – or, to be more accurate, if they were sensible they relied on the expert advice of the engineer. The implementation of those decisions was the secretary's responsibility, and he was left to get on with it. The Committee were, after all, part-time amateur administrators and canal constructors with their own careers to attend to and their own affairs to manage. While they might be tempted to meddle in the glamorous activities of the engineer, they were unlikely to feel any great envy for the life of the administrator or feel any compulsion to take over such jobs as letter-writing, price negotiating, or pacifying angry farmers.

We know even less about the canal administrators as individuals than we do about the engineers, though, as they conducted so much of the company's business, we do know quite a lot about their actual work. But it is possible to get at least an inkling of the sort of man that would be chosen as an administrator. The emphasis was first on respectability – the secretary was the company's public face and it was considered essential that the face should be a respectable one. A great deal of emphasis was placed on sobriety.[1] And having got a suitably sober and respectable employee the company would take steps to try to ensure that he stayed that way. The Huddersfield Canal Company had a rule, for example, that 'no person who shall be employed as Book-keeper or overseer for the said Company shall be allowed to keep a publick House or to sale any victuals, or cloathing, to the workmen employed in or about the works of the said Canal'.[2] It is not difficult to see why such a rule should be necessary, for the possibilities of corruption would otherwise have been enormous. There certainly was bribery and corruption among canal workers at all levels,[3] though it is difficult to estimate just how much went on – by its nature, it was a secretive business and unlikely to find its way into the official records. But certainly some companies did operate a 'truck' system of payment, with all the abuses that that system almost inevitably entailed.

The Committee and the staff needed permanent accommodation. An entry in the Warwick and Birmingham Minute Book suggests that their offices were rather more pleasant than the accommodation provided for Mr Exley at the Lune works. They rented a house and set out a list of the alterations that would

[1] See p. 148.
[2] *Huddersfield Canal Company Minute Book*, 5 March 1795.
[3] See pp. 186–7.

have to be made to turn it into a properly appointed Canal Company Office:

> To put a skirting Board round the Room appointed to the use of the Committee and to put an Hearth stone and Chimney piece to the same.
>
> To build a Brewhouse, Privy and Wall, to make the yard and Garden private and entire – To sink a well and put down a pump – to glaze the windows in the room to be occupied as an office.[1]

The 'brewhouse' was not a standard item of company office equipment.

However, having appointed a suitably respectable secretary and having provided him with an office and a suitably sober staff to man it, the Committee were then able to delegate a great deal of the Company's work. Some of the different aspects of the secretary's job were similar to those of the resident engineer.[2] Both men would be involved, for example, in land purchase, both would be expected to do their bit in bringing difficult contractors to heel, and both would be involved in work measurement. The first job which the secretary would tackle, and which was particularly his own, was that of agreeing terms with the contractors before work started. But here, too, he would seek the engineer's advice – particularly the chief engineer's, as he would be able to give an idea of the going rate in different parts of the country. Settling the rate for the job was a long and complicated matter. There were many different factors to be taken into consideration – what type of country was the canal passing through, were materials to be supplied by the Company or by the contractor, how complicated was the work, would there be any side benefits for the contractor, and so on. Faced with so many variables it was obviously easier to let out work in big lots and save on the paper work; hence the deaf ear which greeted the engineers' requests for smaller lots to be issued.

[1] *Warwick and Birmingham Canal Minute Book*, 8 July 1793.
[2] See above, Chapter 11.

Once the contractors were in, the secretary and his staff were responsible for book-keeping – which often turned out to be a somewhat livelier occupation than one would expect. The large contractors had their own book-keepers, and the two rival clerical forces came up with some very different sets of figures. The Lancaster Committee's books, at one point, showed one set of contractors as having been overpaid by £180, while the contractors' own books came out with a balance of £8,000 owing to themselves. The Lancaster Company finally agreed to pay over an extra £1,500 to avoid work coming to a halt.[1]

Money was a major preoccupation of almost everyone concerned with canal construction, and the major part of the worrying fell to the secretary. It was his responsibility to 'call' on the subscribers to provide the extra funds which their shares in the company specified that they should provide. When a call failed, it was then the secretary's job to wheedle or threaten the non-payer until the funds were forthcoming. While he was trying, often with little success, to get the money from the shareholders, the secretary was also being pressed for money for the works – the engineers needed money to buy essential materials and supplies, the contractors needed money to meet their wage bills. As Archibald Millar, the engineer, remarked of the secretary, no doubt accurately: 'Wood, iron, Coals and Money will give Mr Gregson the Dry Belly Ake.' The dry belly ache was an occupational hazard of hard-pressed secretaries. As the work progressed on the canal, their financial problems grew, and another aspect of finance appeared to occupy the secretary's time – the company's income.

The canal engineer worked on a project until the canal was completed and then moved on – the secretary, on the other hand, had a permanent post. When the construction period was ended, he still had the responsibility of looking after the returns from the tolls and traffic of the canal. This was a simple and straightforward arrangement, but, unfortunately for the hard-pressed secretary, the two periods usually overlapped, sometimes by many years. It was quite common when a sufficient length of

[1] *Lancaster Canal Letter Book*. Letter dated 22 February 1795.

canal was navigable for it to be opened to traffic. This provided the company with a double benefit: there was revenue from tolls and cheaper transport for the company's own supplies. The former might be quite a small sum at first, but the savings on transport costs could be considerable:

> The Coals by the Canal are sold here at 10d p Hund Weight. They cost us before (& do now, if they do not come by the Canal) about 13½d Hund Weight – So you see that we have already a considerable saving as Housekeepers. I have no doubt if we get fairly to work this Concern will do well, but it never will answer the desired purpose either to the publick or to the proprietors until we get over Ribble & communicate with the Coal Country.[1]

This was all a great help to a company very short of money, but all meant extra work for the secretary.

The secretary was liable to be involved in any and all canal activities. He spent as much time as the engineer in trying to control errant contractors. He had no direct authority over them, but he did have one weapon – money. As the representative of the Committee, he was able, if necessary, to withhold payment. It was a delicate problem, for he had to balance the possible gains of forcing the contractor to improve the quality of his work, with the possible losses if the contractor decided to retaliate by stopping work altogether. The secretary usually found that his threats were not particularly convincing. This was partly because so many companies began with a fine flurry of optimism, and quite cheerfully paid out advances to the contractors. When the secretary then began to make threatening noises, he would often discover that his threat was meaningless as the contractors were already overpaid. Accountancy, at that time, could not be described as an exact science. Samuel Gregson, in his long fight with the contractors Pinkerton and Murray, was reduced eventually to write a despairing letter to Rennie to ask him to 'devise some method either by fair means or com-

[1] *Lancaster Canal Letter Book*, Gregson to Rennie, 4 March 1798.

pulsion which will enable us to get on'.[1] Neither engineer nor secretary was able to find the perfect solution.

Although the secretary was supposedly concerned with the overall pattern of administration, his work constantly involved him in the minutiae of canal work – every problem seemed to end up on his desk. The Commitee of the Coventry Canal Company passed over to their secretary the task of handling problems of theft from the workings and ordered him to 'take the necessary measure for the prosecution of Francis Woodhouse charged by Mr Robinson with having stolen part of a plank'.[2] (There is, unfortunately, no indication of how a man could steal part of a plank.) On the same day, he was ordered, without any reason being given, to sack part of the work force.

The main problem and the main source of complaints came from outside the company. With hundreds of men spread over miles of previously peaceful and undisturbed countryside, it was only to be expected that there would be problems. The local dignitaries, the tenants, the landowners, the farmers, the mill-owners, all found their way to the canal office to tell their woes to the secretary. He had the double job of passing on the complaints to the engineers or the contractors, with suggestions for avoiding repetitions, and of pacifying the complainants. Pacification usually came in the form of cash – yet another drain on resources – and, where agreement could not be reached privately, a further expenditure of time and money had to go into taking the affair to arbitration. Arbitration was not popular with either side, as the loser would normally have to pay the costs for both parties. It was a gamble, and some cases that came up seem hardly to have warranted taking the chance. In March 1799, for example, a case was taken to the Commissioners for the Montgomeryshire Canal. Richard Pryce was complaining about loss of water to his mill. The company offered compensation of £19.12s, which the Commissioners eventually decided was fair: Pryce was then faced with having to pay the whole of the costs, which reduced his compensation by £7.1*s*.4*d*., almost half. It

[1] *Lancaster Canal Letter Book*. Letter dated 8 September 1794.

[2] *Coventry Canal Minute Book*, 9 May 1769.

seems a great waste of time and effort. As a rule, the company was reasonable about compensation – in the long run it was easier and probably cheaper. The only case recorded by the Montgomeryshire Commissioners where the final figure was at wide variance with the original offer came in 1814: compensation of £83.12.4½ was awarded against an original offer by the company of £21.1.5. An expensive day for the company, who also had to pay the costs.[1] In general, both sides preferred to settle privately.

Another role which the versatile secretary was expected to play was that of personnel manager. Staff problems came in two main categories: complaints about the staff and workers, and complaints by the staff. The former were the more frequent, and most frequent of all were the ones referring to drunkenness and riotous behaviour. If the complaint referred to the workmen, then it was generally accepted as true without further investigation (but usually with justification) and the standard reprimand was sent. It was regarded as an inevitable fact of life that workmen would get drunk. But if complaints referred to a member of either the office staff or the engineering staff, then that was a very different matter. There was no immediate acceptance of the accuracy of the complaint here: 'Whereas several Reports have been Spread reflecting on the character & conduct of the Officers of this Company, ordered that no regard be had to such Reports unless the same are reduced into Writing.'[2] But if the complaint was found to be justified, then the unfortunate 'Officer' would find the treatment he received rather different from that handed out to the Lune engine-man after his four-day binge. The man not only got the sack, he got a sermon to go with it:

> Complaint having been made to the Canal Come of some bad and disorderly conduct in you on Saturday Evening last and being apprehensive that this is not the first . . . they have given me the disagreeable cross of discharging you from their

[1] *Montgomeryshire Canal Commissioners,* 7 March 1799 and 22 September 1814.

[2] *Oxford Canal Committee Book,* 10 August 1770.

Service . . . to keep you longer in their employ would only give you an Opportunity to lead others astray who perhaps influenced by your Example might become bad Servants to the Company as well as bad members of Society.

I hope this business will be an example to yourself as well as to all other Agents of the Company who may rest assured that no Encouragement whatever will be given to them where there are any Appearances of vice – but wherever sobriety and Strict Attention to the duty required from them is found in any Agent, the Committee will never loose sight of the Man, and his preferment will follow in proportion as he makes himself more useful to the Concern.[1]

Complaints from the staff were, occasionally, departmental disputes, which the secretary was expected to sort out; more commonly, they referred to the popular preoccupation – money. Confronted by a request for extra pay, the secretary's attitude has all the hallmarks of his Victorian successor, combining, in equal proportions, paternalism, morality and economics. Here is Samuel Gregson, writing to the young superintendent of cutting, Thomas Fletcher:

It is certainly the intention of the Committee to give you every reasonable Encouragement if you stay in their Employ. But if by Encouragement you mean an immediate advance in your Salary they give the answer as before (*viz.*) that an advance cannot be agreed to.

If I might advise you as a friend, I would say that your future reputation and establishment as a professional man is of more consequence to you than an advance of Salary at present, you are yet very young and have much to learn. If you continue in a work like ours your abilities must improve and it will be much more to your credit to say you have staid during the completion of a great work than to say you were present during the execution of some parts.[2]

[1] *Lancaster Canal Letter Book*. Letter from Gregson to William Porteus, 5 May 1795.
[2] *Ibid.*, 12 August 1796.

The secretary's job was far from being that of a desk-bound office man – he probably spent almost as much time travelling around the workings as the engineering staff did. Samuel Gregson's letters give a picture of all these varied activities – inspecting and measuring work with the engineer, going to see landowners,, settling payments, all interspersed with a few notes about falling off his horse or being confined to his bed with the rheumatics: by no means an easy life, but certainly a varied one. As a result of all these different activities, the secretary gained quite a reasonable working knowledge of engineering practice, which it was always possible he might be called upon to put to use.

The Warwick and Birmingham Company had troubles with their engineers and, like most other companies in the late 1790s, with money. At the end of 1795, the aqueduct over the River Blyth collapsed and the engineer was dismissed. A year later they were still having problems – the engineering works were in a bad way, and they were too short of money to employ a really competent man, even if one was to be had. It was suggested that the works be handed over to a sub-committee but, as the Committee recorded, although it was an attractive proposition, they would 'find it extremely difficult to select any Gentlemen interested in the Canal who are willing and at leisure to give up a part of their time in such an Employment and are capable of forming an accurate Judgement upon the various Works'.[1] By the Spring of 1798, however, it finally dawned on them that their Clerk-Accountant, P. H. Witton, probably knew as much about the business as most people, and decided that an admirable solution to all their problems would be to add the post of engineer to his existing duties – an answer which said more about the Committee's preoccupation with economics than their common sense. So, for the moderate salary of £250 per annum, Witton became Clerk, Accountant and Engineer. It was only a month before the inevitable entry appeared in the minute book:

[1] *Warwick and Birmingham Canal Minute Book*, 9 November 1796.

P.H.W. having stated to this Committee that the business he is engaged is more than he is capable of executing without an assistant in the out-door business.

Resolved that John Hughes be continued as assistant to Mr Witton.[1]

As a rule, however, the secretary stayed in his own domain, and left the engineering to the engineers.

Witton's salary of £250 for his combination of jobs reflects the average salary of secretaries rather than that of engineers. In general, they were on a considerably lower pay scale. Samuel Gregson's salary was also £250 per annum with expenses of 10s 6d a day for 'horse hire and expences when travelling on the business of the Committee'.[2] This is the same as the salary received by the resident engineer's principal assistant, although the secretary's responsibilities are closer to those of the resident engineer himself. But the secretary had one advantage – security. Once the canal was completed, the engineering staff had to look for new jobs, but the secretary stayed where he was. He could look forward to a period of comparative calm and, hopefully, rising wealth as the expensive construction period came to an end and the revenue at last began to roll in.

In general the secretary's main attribute was versatility – there was very little that went on in the canal business in which he could not expect, at some time or other, to find himself involved. The other members of the administrative staff had calmer and less varied lives. The company's solicitor, unless he was involved in the nerve-tearing business of Parliamentary work, would perform the usual legalistic functions – drawing up contracts, giving advice on the many disputes that arose between the company and others, and so on. The junior accountants and clerks had their own, rather dull, jobs to attend to. The only excitement which could be expected to come their way would come on pay-days, when they would take the money to the contractors. It was not so much the fear of robbery that caused the

[1] *Ibid.*, 22 May 1798.
[2] *Lancaster Canal Committee Minute Book*, 3 July 1792.

excitement, though wage robberies were by no means unknown, as the situation that could and did arise when they had to face a crowd of navvies and explain exactly why there would be delays in the payment of wages. The reaction of the navvies in those circumstances was all too predictable.

PART THREE
THE WORKERS

13

THE NAVVY

A stranger called Thomas Jones supposed from Shropshire, having been unfortunately killed in the works near Gannow by a fall of earth.

Entry in *Leeds Committee Minute Book*, Leeds and Liverpool Canal Company, 27 February 1800.

Of all the men concerned in canal building, the workmen retain the greatest degree of anonymity: even when they died in the works, the most that could be said of them was that they were strangers of uncertain origin. And yet, in many ways, they are the most interesting of them all. The story of the canal worker is infuriatingly incomplete and discontinuous – we can pick up a few facts from the beginning of the period, a few more from the middle and a fair number from the end. The gaps have to be filled in by speculation.

For well over a century the navvy was an important figure in the social and economic history of Britain. It was tens of thousands of navvies who, with the simplest of tools – spade and barrow – dug their way from one end of the country to the other, and, without their skill and trained muscles to form the basis for the railway work force, that great system could never have grown as it did. It was the canal 'navigator' who became transformed into the railway 'navvy'.

The first difficulty in trying to determine facts about the canal navvy and his life is that everyone ignored him, except for the occasions when he was off on one of his periodic bursts of rioting.

He was very like Chesterton's 'invisible' postman; he was such an accepted part of the scene that the writers and travellers who went to look at the work in progress rarely noticed his presence or felt it worthy of mention. If he was noticed at all, it was not as an individual, nor even as one of a particularly important group; he was far more likely to end up as the subject of a literary image or a classical analogy. A tourist's comment on the Bridgewater construction is typical:

> I surveyed the duke's men for two hours, and think the industry of bees, or labour of ants, is not to be compared with them. Each man's work seemed to depend, and be connected with his neighbour's, and the whole posse appeared, as I conceive did that of the Tyrians, when they wanted houses to put their heads in, and were building Carthage.[1]

It seems highly unlikely that a gang of Lancashire workmen did, in fact, have very much in common with the builders of Carthage. Even Robert Southey, one of the most observant and accurate recorders of canal construction, could do no better than to spare a few words for the workers at the diggings – 'emmets to an ant-hill' was the sum total of his comment. So we have to look elsewhere to find where the navvies came from and how the navvy working force grew into the mobile, tough and skilled group at the beginning of the nineteenth century.

The beginning, again, has to be the Bridgewater Canal. Even in those early days of canal building, there were still as many as six hundred men employed in the diggings, split up into gangs of fifty, each with its own foreman. These were locals, many of them recruited from the Duke's own estates and mines. Smiles' account gives a good idea of the sort of men they were:

> Brindley did not want for good men to carry out his plans. He found plenty of labourers in the neighbourhood accustomed to hard work, who speedily became expert excavators; and though there was at first a lack of skilled carpenters, blacksmiths, and bricklayers, they soon became

[1] Quoted in Anon., *A History of Inland Navigations* (1779).

trained into such under the vigilant eye of so able a master as Brindley was. We find him, in his notebook, often referring to the men by their names, or rather byenames; for in Lancashire proper names seem to have been little used at that time. 'Black David' was one of the foremen employed on difficult matters, and 'Bill o Toms' and 'Busick Jack' seem also to have been confidential workmen in their respective departments. We are informed by a gentleman of the neighbourhood that most of the labourers employed were of a superior class, and some of them were 'wise' or 'cunning men', blood-stoppers, herb-doctors, and planet-rulers, such as are still to be found in the neighbourhood of Manchester. Their very superstitions, says our informant, made them thinkers and calculators. The foreman bricklayer, for instance, as his son used afterwards to relate, always 'ruled the planets to find out the lucky days on which to commence any important work,' and he added, 'none of our work ever gave way'. The skilled men had their trade secrets, in which the unskilled were duly initiated, – simple matters in themselves, but not without their uses.[1]

This is still a long way from the 'navvy'. But in those early days, canal work was merely a part-time job, a change from harvesting or mining. The men who came to work on the Bridgewater probably had no idea of continuing in canal work once the waterway was finished. They were simply untrained labourers – but then there were no trained canal labourers, nor was there any recognized training. When a new canal concern was started, men learned what they could about the techniques by example. The Coventry Canal was begun in 1768 and the more skilled workers had to be sent off to other works to find out how things were done:

Ordered at the recommendation of Mr Brindley that Mr Bull one of our clerks do go for his Improvement to the Staffordshire and other Navigations now forming for a Time not exceeding Three weeks.

[1] Samuel Smiles, *op. cit.*, Vol. 1.

> Ordered that John Paish now employed by us as a Carpenter be sent for his Improvement to the Birmingham Navigation and there remain during the building of one of the Locks to inform himself fully of the Nature and Construction thereof.[1]

Everything had to be learned, even such an apparently simple matter as building a wheel-barrow: 'Resolved that one hundred wheelbarrows be provided and that an advertisement be published for Persons to undertake the making thereof according to the Model lately sent from Staffordshire.'[2]

Slowly a body of experienced canal workers began to form. It might seem that there was not a great deal of skill needed in shovelling earth out of the ground into a barrow and wheeling it away, but the digger learned technique and, as his muscles were trained to the work, he became stronger. It was estimated that, in good soil, an experienced navvy could shift twelve cubic yards of earth a day.[3] Anyone who has ever spent a Sunday afternoon digging the back garden should be able to appreciate the effort involved in digging a trench 3 feet wide, 3 feet deep and 36 feet long. In a period of poverty, the man who had this ability, and was prepared to stand the gruelling life, would find himself one of the few workers whose services were constantly in demand. Workers therefore began to travel to exploit their new-found skills. Groups of men would set off as soon as one set of diggings was closed to look for the next. The days of extensive local labour recruitment were numbered. When the Reverend Shaw visited the workings at Greywell Tunnel on the Basingstoke Canal in about 1780, the transition was virtually complete:

> I . . . saw about 100 men at work, preparing a wide passage for the approach to the mouth, but they had not entered the hill. The morning was remarkably fine, 'The pale descending year, yet pleasing still,' and such an assembly of these sons of labour greatly enlivened the scene. The contractor, agreeable to the request of the company of proprietors, gives the prefer-

[1] *Coventry Canal Minute Book*, 6 June 1769.
[2] *Ibid.*, 9 March 1768.
[3] W. Tatham, *The Political Economy of Inland Navigation* (1799).

ence to all the natives who are desirous of this work, but such is the power of use over nature, that while these industrious poor are by all their efforts incapable of earning a sustenance, those who are brought from similar works, cheerfully obtain a comfortable support.[1]

The experience of the regular navvy was beginning to tell. It is impossible to state any particular date when the change-over became complete; it was a gradual process and there were probably many local variations. It seems fairly certain that by the 1790s the system of contractors with their gangs, and individual workmen, touring the country was firmly established. Certainly, by the beginning of the nineteenth century the system was so set that Thomas Telford, who deliberately set out to reverse the trend for the Caledonian Canal workings, had to publish detailed figures to convince people that he was not using English and Irish labourers in any great numbers. But even Telford, who wanted to train the Highlanders to the work, found it necessary to bring in experienced workers:

> Several people of the Highlands who have been engaged in Canal-work in other parts of Scotland and in England, have begun to work: and they may be expected to prove useful examples to others who have not been accustomed to that sort of employment.[2]

Telford managed to keep to his policy of training and using local labour and, in any case, the Caledonian offered fewer attractions than other canal works. It was remote, which meant that a man taken on by the Caledonian would not find it easy to move on again if he felt like it, and, more importantly, the pay was comparatively poor, which discouraged those who took the trouble to inquire about work: 'The Rate of Wages hitherto given to the labourers upon these works, has been about Eighteen-pence a day . . . and the numbers of Labourers who

[1] Rev. S. Shaw, *op. cit.*
[2] *First Report of the Committee for Making and Maintaining the Caledonian Canal*, 1804.

have from time to time offered themselves for work, and demanded a higher Rate, have been uniformly rejected.'[1]

By this time the pattern was very clear – the navvy was now a roving, migrant worker, hiring his labour where work was to be found at the highest rate of pay. His wandering habits made life difficult for the engineers:

> The works on the upper Level go on slowly principally owing to the Men being enticed to leave the works by greater advantage being held out to them by a Contractor at Bristol. But I have some expectation that some of them will return after finding their Disappointment. The Consequence will be that the Contractor must pay higher wages.[2]

The navvy had arrived.

The obvious question, and the most difficult to answer, is where the navvies came from. The first recruits came into canal work almost by accident – they were available when the canal came to their neighbourhood. Many of them must have stayed in their own neighbourhood when the work finished, but a few seem to have stayed with the canals. There were men from the north of England and the Midlands and a few who came from the fens, the bankers who were already experienced in this sort of work. These formed the nucleus, but the later recruits to the travelling groups came mainly from the most depressed areas of Britain – Scotland and, above all, Ireland.

In the 1790s, the demand for canal labour grew at a rate which meant that it was impossible for the supply of experienced labour to keep pace with the demand. Up to the start of the decade, Parliament had authorized construction of some twenty canals, some of which were still incomplete. In the next five years, from 1790 to 1795, more than twice that number was authorized.[3] There are no accurate figures, but, at a conservative estimate, there must have been well over 50,000 labourers em-

[1] *Ibid.*
[2] *Engineers' Reports to the Management Committee, Kennet and Avon Canal Company*, 16 July 1804.
[3] See *Chronology*, pp. 219–23.

ployed on canal work by the end of the century. Other users of labour became alarmed by this apparently insatiable demand for workers, and the farmers tried to persuade Parliament to put a stop to it. Sir Charles Morgan moved for leave to bring in a Bill to 'restrain the employment of labourers in the time of the corn harvest'. The debates on the motion are interesting, partly for the indications they give of the farmers' attitudes towards their labourers and partly for the evidence they supply about the sources of labour for the canals.

The farmers' claim was simple: workers were being lured away from the land by the prospect of better pay on the canals. This, they felt, was ruinous to the prospects of getting in the harvest on time. The Bill's opponents pointed out that, in effect, the farmers were asking Parliament to force workers to do work which they preferred not to do at a rate which was less than they could get elsewhere. Why could the farmers not pay a decent wage instead of asking for compulsory labour? A number of speakers pointed out that many of the labourers had, in any case, never had any connection with farming or harvesting:

> Mr Dent was against the bill in question; he said there were hundreds of people who came from Scotland and from Ireland, for the purpose only of working in canals, and who knew nothing of corn harvest. If this bill passed, they would be entirely deprived of the only honest means they had of subsistence.[1]
>
> Mr Courtenay spoke strongly against the arbitrary principles of this Bill. . . An Hon. Gentleman had stated a difficulty with regard to Scotland, and there was another with regard to Ireland. The people from that country, who worked in bogs, knew nothing about the corn harvest, and to send them to work at the harvest was to make them starve, or to turn highwaymen and robbers.[2]

The Bill never reached the Statute books and Sir Charles

[1] *Morning Chronicle*, 11 April 1793.
[2] *The Times*, 11 April 1793.

Morgan was left complaining that 'he despaired of getting in the corn'. The arguments used in the debate reinforce the idea that, at that time, there was a mixture of English workers, possibly recruited from the farms, and a specialized work force that had come from Scotland and Ireland specifically to work on the canals. Why did workers travel many miles from their homes to take on the rough, hard, nomadic life of the navvy? In the case of the Irish, the answer is plain enough.

The Irish already had a tradition of coming to England to find work, often, in spite of Mr Courtenay's argument, on the farms.[1] When the demand for canal workers began to grow so rapidly, it became common for the contractors to advertise in Ireland. The cost of getting to England was not too high, if you were not too fussy about how you travelled: the rates in 1779 were: 'For cabin passage, with luggage, £1.1s; for steerage, 10s.6d; for the hold, in which the poor labourer with his wallet, on which he rests his weary head, 2s.6d.'[2] They may have had to suffer an uncomfortable journey, but almost anything they met on arrival was likely to be better than the conditions they left behind. The Parliamentary Committee on Emigration from the United Kingdom heard evidence on the state of the Irish poor in the early part of the nineteenth century. No further comment is necessary:

> Are they [the huts they live in] not built upon the bog itself sometimes? – In many instances the very bog.
> Upon the mere bog sod? – Yes.
> Is not the roof formed with a few sticks? – Yes, some sticks thrown across.
> Without straw? – Yes, but with bog sods.
> What is the nature of the furniture inside one of these huts? – They generally have a pot and a little crock, and very few other articles.
> . . . .
> What do they sleep upon? – Very often rushes and straw.

[1] See p. 16.
[2] Thomas Baines, *History of the Commerce and Town of Liverpool* (1852).

Are these habitations divided into apartments of any kind? – Generally in one; there may be one little partition.
What sort of bed clothes have they? – O, very bad; their clothing is all very bad.

. . . .

What is the food of this lowest class of labourers? – Potatoes, nothing else.
What do they drink with them? – In summer some of them get a little butter milk, in the winter seldom any thing but the salt and the water.[1]

The Irish were the poorest, but the Highland Scots were very little better off. A great deal of the Committee on Emigration Report is taken up with the conditions prevailing in Scotland. What appears is a depressed population, in every sense, who saw their only hope in emigration. In some cases, men would go into England to find work in such projects as canal construction – a great number went much farther to start a new life in the overseas colonies. Telford, like many other Scots, was seriously concerned by this depopulation, and frequently emphasized the usefulness of the Caledonian in helping the local unemployed: 'there is reasonable expectation that employment may be found for many Hundreds of Persons, who at present have no means of employing themselves in profitable labour, and who may otherwise be more readily induced to emigrate'.[2]

In England and Wales, conditions varied between different regions and between different times. Many canal labourers were recruited from the farms – the farm labourer and the small tenant farmer. The small farmer had always been in a peculiarly vulnerable position, at the mercy of the weather. In the eighteenth century his position became markedly more difficult, as two factors pressed in on him. First, there was the spread of enclosures, which reduced the amount of common land available; and then there was the growth of industry. The industrial

[1] Evidence of Hugh Dixon, Westmeath landowner, *Third Report from the Select Committee on Emigration from the United Kingdom*, 1827.
[2] *First Report of the Caledonian Committee*, 1804.

development almost certainly had the most profound effect, for many small farmers and labourers had relied on 'cottage industries' – weaving and spinning – to supplement their income and keep them and their families above subsistence level. The factory system changed all that – the demand for hand work fell and many were reduced to appalling poverty. The number of labourers, as opposed to independent farmers, grew. They could seldom get jobs in factories, which preferred to employ women and children, and farm work was scarce, except at harvest time. The canals offered work; and the labourers moved in their tens of thousands across the face of the country. Between them they dug almost three thousand miles of waterways.

There were many commentators to eulogize the engineers who designed the canal system, but there were few who could find a good word for the men who actually dug it. On the other hand, there were quite a few to record their misdeeds. Take thousands of poor, uneducated men, remove them from home and family, send them out into the wilds to sweat away at hard, dirty and dangerous work, and you cannot be too surprised if the end product is a gang of men who frequently find their release in outbursts of drunkenness and fighting.

> In the making of canals, it is the general custom to employ gangs of hands who travel from one work to another and do nothing else.
>
> These banditti, known in some parts of England by the name of 'Navies' or 'Navigators', and in others by that of 'Bankers', are generally the terror of the surrounding country; they are as completely a class by themselves as the Gipsies. Possessed of all the daring recklessness of the Smuggler, without any of his redeeming qualities, their ferocious behaviour can only be equalled by the brutality of their language. It may be truly said, their hand is against every man, and before they have been long located, every man's hand is against them; and woe befal any woman, with the slightest share of modesty, whose ears they can assail.
>
> From being long known to each other, they in general act

in concert, and put at defiance any local constabulary force; consequently crimes of the most atrocious character are common, and robbery, without an attempt at concealment, has been an everyday occurrence, wherever they have been congregated in large numbers. . . .[1]

This account is almost certainly exaggerated. Lecount was engaging in a spot of special pleading, for he was arguing against the adoption of the same system for the railways. But the newspapers of the eighteenth century and the canal company records have enough accounts to show that his view of the navvy was not by any means uncommon or too far from the truth.

The Sampford Peverell riot of 1811 was typical of scores of others, and began with an argument about pay. Many contractors paid their men in tokens, which could then be exchanged for goods or, theoretically, cash. In some cases, however, if there was any doubt about the solvency or reliability of the company issuing the tokens, the navvies had a great deal of difficulty in getting anyone to cash them. Their limited patience soon ran out:

On Monday last a disturbance of a serious nature occurred at Sampford Peverell. The annual fair for the sale of cattle, &c was held there on that day. On the Saturday preceding, a number of the workmen, employed in excavating the bed of the Grand Western Canal, assembled at Wellington, for the purpose of obtaining change for the payment of their wages, which there has been lately considerable difficulty in procuring. Many of them indulged in inordinate drinking and committed various excesses at Tiverton, and other places, to which they had gone for the purpose above stated. On Monday the fair at Sampford seemed to afford a welcome opportunity for the gratification of their tumultuary disposition. Much rioting took place in the course of the day, and towards evening a body of these men, consisting of not less than 300, had assembled in the village. Mr Chave (whose name we had

[1] Peter Lecount, *The History of the Railways connecting London and Birmingham* (1839).

occasion to mention in unravelling the imposture respecting the Sampford ghost) was met on the road and recognized by some of the party. Opprobrious language was applied to him, but whether on that subject, or not, we have not been informed. The rioters followed him to the house, the windows of which they broke; and, apprehensive of further violence, Mr Chave considered it necessary to his defence to discharge a loaded pistol at the assailants. This unfortunately took effect, and one man fell dead on the spot. A pistol was also fired by a person within the house, which so severely wounded another man, that his life is despaired of. A carter, employed by Mr Chave, was most dreadfully beaten by the mob. Additional members were accumulating when our accounts were sent off, and we understand their determination was to pull down the house.[1]

The authorities did not seem overkeen to tangle with the rampaging navvies, as the local paper indignantly noted:

It is impossible not to feel the deepest abhorrence for the proceedings of a savage ungovernable banditti, whose ferocious behaviour we hope will be visited by the heaviest punishments of the Law. . . . It is a most extraordinary circumstance that the whole village and neighbourhood should have been kept in a state of the greatest terror and commotion for more than twenty-four hours, and no effects of the Police or Military made to quell the tumult. In the name of Justice, where are the Magistrates![2]

If the navvies sound, from this account, to be a peculiarly riotous and bloodthirsty mob, it is worth looking at their actions in relation to the whole period. It was a time marked by

[1] *Annual Register*, April 1811. The 'Sampford Ghost' was an invention of Mr Chave's. He had taken a shop in the district and had used the ghost to get rid of an unwanted tenant. The phantom's habits included rattling chains and beating out the rhythms of 'Go to Bed Tom' in the room above the tenant's head. The interest aroused had brought new customers to come and see the haunted shop – until the hoax had been exposed.
[2] *Taunton Courier*, 25 April 1811.

25 Rebuilding lock No. 24 on the Old Union Canal (now part of the Grand Union) in 1896. Almost nothing has changed as far as techniques are concerned: most of the navvies' tools shown in the title pages can be seen in this photograph.

26 The Islington tunnel on the Regent's Canal under construction. The deep cutting up to the mouth has been completed.

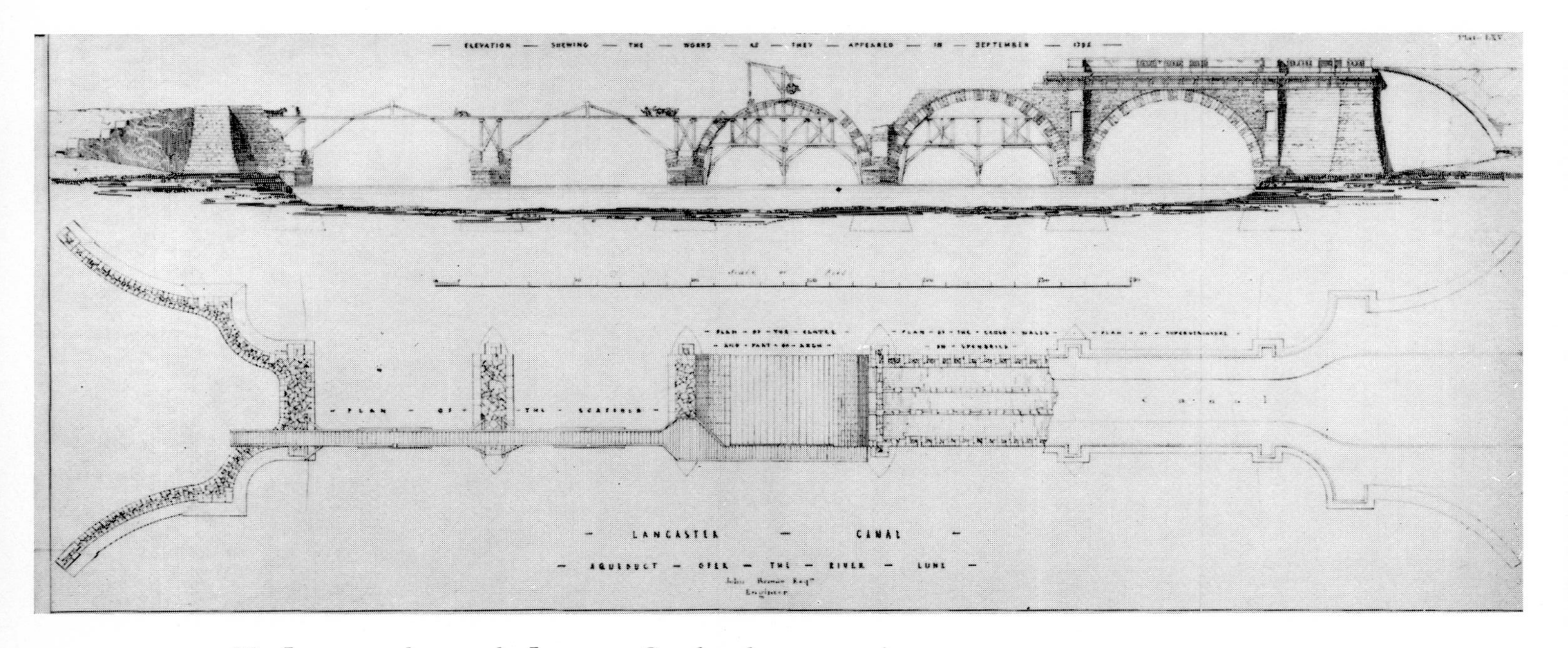

27 The Lune aqueduct on the Lancaster Canal under construction.

recurrent rioting, mostly over the price and supply of bread. In August 1795, there was a particularly serious riot at Barrow-upon-Soar in Leicestershire, which ended with the military opening fire, killing three of the rioters and seriously wounding eight others. This was only one of a series of bread riots, and pamphlets were distributed in many parts of the country which helped to rouse an already angry people:

> Treason! Treason! Treason! Against the People!
> The People's humbugg'd! A plot is discovered. Pitt and the Committee for Bread are combined Together to starve the poor into the army and the navy and to starve your widows and orphans! . . . Sharpen your weapons and spare not! for all the scrats in the nations are united against your Blood! Your wives and your little ones![1]

In this riotous age, the notoriety of the navvies made them admirable scapegoats. It was far more convenient for the authorities to put the blame on the strangers than to admit that there might be a good reason for the behaviour of the poor:

> The brown bread was very good, and the unquietude of the populace, from an idea of scarcity, is far less than where riots have been set forth by the news-writers, in some places most erroneously, the disturbance at Barrow on Soar excepted, which has indeed been productive of the most fatal consequences; but this, it should be recollected, was among that newly-created, and so wantonly multiplied, set of men, the diggers and conductors of navigations.[2]

But even if the last writer's allegations are untrue, for them to have sounded at all reasonable they must have fitted into an accepted picture of navvy behaviour. It certainly is more than likely that the navvies took part in the Barrow riot, even if they did not actually start it. There was a strong tradition among the navvies of never refusing a fight if one was offered, and never staying out of someone else's fight if they happened to be around

[1] Quoted in *The Oracle*, 12 August 1795.
[2] *Gentleman's Magazine*, August 1795.

at the time. There are many stories in the localities where the canal builders came of fights between bands of navvies and local workers. There was, for example, almost a full-scale battle between the navvies and the needle-makers of the Midlands, where the two sides took advantage of the enclosure of some near-by land to uproot the fences and use the posts as clubs.

The navvies could be sure, when they came to a new district, that there would be more to curse them than to welcome them. A Lancashire clergyman preached a sermon on the subject, though there is no record that any navvies came to listen. He began with a half-hearted welcome to the newcomers, then got down to the main business of a tirade against the navvies, who 'cheat, and steal, and drink, and swear, and fight, and do all kinds of mischief to themselves and others'.[1] Their reputation, in fact, had grown to such fearsome proportions that the canal navvies do actually appear to have been the first genuine example of 'Rent-a-Crowd'. Henry Eastburn reported to the Lancaster Canal Committee:

> We are going on very ill with work in this neighbourhood, not a Man has been at work since the Canvassing began & I doubt it will be the case as long as the Election continues. Lord Stanley's party having hired them, which can be for no other purpose than to riot & do mischief.[2]

A comment which tells us as much about eighteenth-century politics as it does about navvies.

The majority of reports of rioting come from the 1790s and later, which reinforces the suggestion that the gangs of navvies, moving in bands across the country, became the norm among canal workers at about that time. While most people were vociferous in welcoming the idea of a canal being built in their district, only the inn-keepers showed any signs of welcoming the men who came to build it. They were assured of a large and rapid increase in trade, though many of them were not satisfied with that.

[1] *Canal Boatman's Magazine*, Vol. 1, No. 1, 1829.
[2] *Mr Eastburn's Reports*, 30 May 1796.

> The Labourers for cutting the canal are much imposed on by extravigant charges of the Inn keepers and the Committee are desired to consider if any scheme can be come into for the convenience of such Labourers by erecting Tents, Booths, &c &c providing them with meat & drink at a more easy expence.[1]

The relationship between the navvies and the communities near which they worked was not as bad at first as it became later, and was partly dependent on where the navvies actually lived. In the early days, when labour was almost all recruited locally, the navvies remained a recognizable section of the local community and there were few problems. Even when the workings progressed and a worker would have a long distance to travel to get to the diggings, he was still able to visit his home regularly. The trouble came with men who had completely cut themselves off from home and family – the migrant navvies. Canal companies tried, wherever possible, to get the men into lodging houses, on the grounds that they provided a link, even if a tenuous one, with a normal life. But as canals became more ambitious, the numbers of men required went up, and accommodation in lodgings was no longer possible. When that stage was reached, the men had to live in special accommodation built at the workings. Hundreds of them would be gathered together in one place – even in the very early days, a project such as the Harecastle was employing as many as 600 men.

A particularly useful source of information about the living conditions of workmen on the line is the set of Reports of the Caledonian Canal Committee. They show how the numbers involved fluctuated with the work and with the fortunes of the surrounding regions. In 1805, when work was begun, there were only 150 men employed. Most of the labourers came from Argyll and Inverness – the exceptions being the experienced workmen specially brought in. Over the next few years, the numbers fluctuated between 150 and 900, as men disappeared to go off to the harvest or to take seasonal work in the herring fisheries. By 1812 the numbers had risen to 'upwards of

[1] *Leeds and Liverpool Canal Committee Minute Book*, 3 January 1771.

Fourteen Hundred' and the Report explains the increase as being due to 'the temporary depression of trade at Glasgow, which has thrown Seven Hundred Masons and many other Workmen out of employment'. Such large numbers could not be accommodated in what was an almost unpopulated area, so accommodation had to be provided:

> The accommodation and markets near the intended Canal, and especially at the South-West End of it, are so little adequate to the wants of a numerous body of Workmen and Labourers, that we have found it necessary to continue our attention to their habitations and subsistence; temporary Sheds and Huts for lodging most of the Labourers have therefore been erected; and we have continued in some degree the supply of Oat meal at prime cost. . . . With a further view to the welfare of the Persons employed, We have encouraged the establishment of a small Brewery at Corpach, that the Workmen may be induced to relinquish the pernicious habit of drinking Whiskey; and cows are kept at the same place to supply them with Milk on reasonable terms.[1]

One wonders how successful the Committee were in weaning a thousand hard-working Scots from their whisky. The accommodation could not by any stretch of the imagination be described as luxurious. The huts and sheds were made out of turf, and for these the workers had to pay rent to the Company. Later, when buildings such as lock-cottages and stables were erected along the canal, these were used to provide accommodation for the navvies – though they had to share their living space with 'the Horses employed on the different Railways'. The food was no more exciting than the accommodation – oatmeal and potatoes.

The living conditions of the workers on the Caledonian were probably very much the same as those found on other construction sites – certainly the workers who came up from England were not in the least deterred by the conditions they found. They might have a poor opinion of the rates of pay, and there

[1] *Second Report*, 17 May 1805.

were disturbances on this score, but no complaints about the conditions were recorded. If this is a fair picture of the life of the navvy – a turf hovel and dull, plain food is all he had to come back to at the end of work – then there is small wonder that men went wild on the few occasions when they could get away. They worked hard for the little they got, and owed allegiance to no one but the contractor who hired them.

14

THE CONTRACTORS

To Bricklayers, Masons, Canal Cutters etc – Notice is hereby given, that all Persons disposed to contract for excavating the Entrance Basin near the Thames at Rotherhithe, performing the Masonry, Brick-work &c of the Walls, and constructing the necessary Locks, may deliver sealed Proposals for the same, addressed to the Committee at their Office, Will's Coffee-House, Cornhill, on Friday the 29th inst. . . . Security will be required for the due execution of the Contract.

Advertisement placed by the Grand Surrey Canal Company, *The Times*, 8 July 1803.

The contractors formed the link between the company and the workers. The navvies themselves had very little to do with either the company or the company's officers: it was the contractors who hired them, the contractors who paid them, and the contractors who fired them.

The system of letting out construction work to independent contractors was established at the beginning of the canal building period, although on the first projects the system was incomplete, with the company still being directly in control of men and equipment. The Coventry Canal, for example, was begun in 1768 and the contractors were only expected to provide labourers for the straightforward jobs of cutting and puddling. The company provided all the equipment – planks, shovels, picks and wheelbarrows. This turned out to be an unsatisfactory arrangement. For a start, there was the expense – the wheelbarrows, for example, were specially made for the company by John Roberts of Whitherley who agreed to produce a hundred barrows at seven shillings each and deliver them at the rate of twenty a month. As they were the company's property, they had to pay someone to keep an eye on them – 'One man 2 Day watching Barrows 0-2-0' – and then had to pay more men to move them from one part of the works to another – 'John Armstrong 6 days looking over odd men & gathering barrows & tools'.[1] It all took time and money, and even then they lost them: 'Ordered that an advertisement be published in Hand Bills and in the Coventry News Paper offering a Reward of Five Guineas for the Discovery of any Person who has stolen any of the Rails belonging to the Company or taken away Wheelbarrows or other Utensils.'[2] Eventually, it became obvious that the system was inefficient, and later contracts all specified that the contractors would have to supply their own equipment.

The contracts for digging were, at first, modest. No particular skills were looked for – all the contractor was asked to do was to provide enough able-bodied men who were willing to spend all day digging out a canal. The cash involved came in equally modest amounts: in the first stages there were nine contractors at work, sharing out a monthly total of £180, but a year later there were fourteen contractors and the cash involved had risen to almost £1,500.[3] But, at this time, the company was still employing direct labour; skilled workers such as carpenters and masons were brought in by the company to look after the specialized jobs – lock construction and bridge building. The company also tried to keep some sort of control over the labourers at work in the diggings: 'Ordered that Mr Wheeler do super-intend the Labourers . . . and take check at proper times of their Numbers and he is impowered to discharge any of them who are negligent.'[4] This system too was abandoned as it became clear that the company was getting neither the advantages of

[1] *Coventry Canal Journal*, 24 September and 26 November 1768.
[2] *Coventry Canal Minute Book*, 29 August 1769.
[3] *Coventry Canal Journal*, 28 May 1768 and 29 April 1769.
[4] *Coventry Canal Minute Book*, 14 November 1769.

employing only direct labour nor the advantages of letting out all the work. In the end, they settled for the latter course.

The contracting system changed over the years, and the contractors changed with the system. At first most contracting firms were small concerns – either a group of men willing to take on cutting work, or local builders and masons who contracted for locks and bridges. Here are two typical contracts for the period, which were agreed by the Clerk of Works for the Oxford Canal Company in 1770: with John Watts 'for compleating to Mr Brindley's satisfaction the Canal and Towing Path thro' the whole extent of Lord Craven's Estate at the rate of £350 per mile . . . also for building two Wagon Bridges of good Brick coped with stone at the price of Two Hundred and Ten Pounds for both'; and with John Robinson 'for all the level cutting from the Lane on the East Side of Anstey to the Valley betwixt Withybrooke and Combe Fields at three pence three farthings per cubic yard for digging and for finishing the slopes and back slopes and laying the soil again'.[1]

The demand for labour grew steadily, and the companies were no longer satisfied with employing local, untrained and unskilled labourers – they wanted trained navvies and experienced contractors. The canal companies were prepared to go much further afield and no longer relied on local resources. When, for instance, the Lancaster Company wanted to let out work they listed the newspapers in which they intended to advertise: 'Liverpool, Leeds, Newcastle, Glasgow, Birmingham, Manchester, Whitehaven, Derby, Edinburgh, Gen' Evening, St James Chronicle.'[2] They were no longer dealing with local contractors, but with contractors who were sending their men all over the country. So we find the same contractor in canal workings as far apart as Lancashire and Wiltshire. The scale of the contracts changed as well – there is a world of difference between the £210 contract for bridges of 1770 and the contracts given to Mr Pinkerton of Oldham and Mr John Murray of Colne by the Lancaster Canal Company for 'cutting, banking

[1] *Oxford Canal Committee Minute Book*, 2 January 1770.
[2] *Lancaster Canal Committee Minute Book*, 23 October 1792.

and completing' and 'work on aqueducts, bridges etc.' which was valued at £52,000.[1]

Although by the end of the canal building period the major part of the work was in the hands of the big contractors, employing hundreds of men, the smaller concerns still found a place within the system. Sometimes they were able to get a small piece of the work for themselves; more often they worked as subcontractors, helping out the big concerns whose resources were overstretched.

Where did the contractors come from? The majority simply grew up with the canals – a demand was created, and it was in the nature of things that someone should come along to fill the need. The first contractors were hardly more than spokesmen and negotiators for a group of labourers. Many contracts, sometimes involving sums reaching into thousands of pounds, were made with men who could only sign their names with a cross, and who were described simply as 'canal cutters' or 'diggers' and who quite often had no fixed address. As time went on, it became more difficult for a man to get started in this way. It was no longer enough to supply the men, he was expected to supply materials as well. William Montague, for example, was a canal cutter who moved from Berkshire to Gloucester to take a contract on the Gloucester and Berkeley Canal. It specified that 'the said William Montague shall at his own cost and charge find and provide all Machines, Planks, Wheelbarrows, Trussells, Gang Ladders and all other Tools and Implements'. In the case of the contracts for tunnelling on the Kennet and Avon Canal it was even specified that the contractors should provide their own horse gins and steam engines. This all required capital, and it also became usual for the company to insist on a bond being deposited. John Pixton, who also contracted with the Gloucester and Berkeley, was held responsible for any damage that might be caused by his men and had to deposit £200, which would be forfeited if he left before the work was finished. Occasionally, a Company would agree to loan equipment, but this was certainly not usual practice. If a navvy wanted to start up as a contractor,

[1] *Ibid.*, 31 December 1792.

then he had to find equipment and money from somewhere. One answer might be a small consortium: a contract for building an embankment on the Brecnock and Abergavenny Canal went to a group consisting of William Watkins, labourer, Thomas Powell, shopkeeper, and William Parry, gentleman – experience, equipment and cash. Sometimes the company's own employees left to set up as contractors – once a supervisor had enough experience there was a strong temptation to cash in on it. Contractors' employees, too, sometimes set out on their own. Many big contractors never visited the works – to the extreme annoyance of the engineers – and left the entire operation under the control of a foreman. The foreman might quite reasonably decide that if he was going to do all the work, he might as well see some of the money.

The big contractors usually had a fixed headquarters, conveniently situated for travelling to the different canal works. John Murray, for example, originally came from the Glasgow area, but settled in Lancashire. If, however, they became really big, then they no longer even considered such factors. An indication of just how far the business of contracting for canal work grew can be had by comparing the modest contracts of the wandering cutters with the contract given for completing the work on the Gloucester and Berkeley Canal. It went in 1823 to 'Hugh McIntosh of Bloomsbury Square – Contractor for Public Works', and was worth £111,493.15*s*.11*d*.

From the entries in canal records, the impression given by the contractors is of a group of men who, at their best, were quarrelsome and incompetent and, at their worst, rogues and thieves. But this is a very one-sided view. For one thing, it was only the complaints that found their way into the records and, if there was a dispute, the company was unlikely to put down any side of the argument except their own. Undoubtedly, many contractors simply took a contract, did their work satisfactorily, collected their money and moved on. But, even so, there are so many disputes recorded that they must have represented a considerable problem to the companies. The Lancaster Canal Company certainly had more than its fair share of troubles with contractors, especially with Pinkerton and Murray. The story shows up the disadvantages of the whole system. Basically, the two sides had conflicting interests: the company wanted good workmanship, carefully done; the contractors wanted to cover the greatest possible distance in the shortest time and at the least cost to themselves.

Work on the Lancaster began in January 1793 and at first everything went well: 'We are going on very briskly with the cutting & labourers crowd in'.[1] The cutting continued to go well, but the masonry work, which had been let to Pinkerton was not in quite such a happy state: 'The cutters are now spread from Lune to Conder & get on very fast, but we have not yet a single stone carted to any part of the Work. The Mason Work we very much fear will be tedious and hinder the cutting.'[2] That was the start of the trouble. By June, relations between the company and Pinkerton and Murray were becoming decidedly strained. The Secretary, Samuel Gregson, wrote to Rennie:

> Mr Murray is wanting to begin upon the New Lot, but the Come have refused to give him liberty at present because he may employ a greater number of Men on the present Lot, & the Land on the New Lot is not agreed for. The Come have great complaints that the work is not finished off as they go along, and a number of more hands might be employed without rambling over the Country & creating Damage unnecessarily.[3]

By the end of the first year, disillusionment was total, and the tone of the letters to the two contractors had become almost despairing:

> Messrs Pinkerton & Murray
>
> . . . In respect to the Quarry at Cross Hill you ought to have known whether it would answer your works or not

[1] *Lancaster Canal Letter Book.* Letter from Samuel Gregson, 27 January 1793.
[2] *Ibid.*, 2 February 1793.
[3] *Ibid.*, 5 June 1793.

before you had gone 'to the great expence' as you state, and it will be well for you to fix on the place you propose for the Quarry to answer for your works before you apply for a road and also to give in an estimate of the expence, as the Come do not think it reasonable or right that they should make a road from every place that you choose to pick a few stones out of....

The Come are sorry that they have reason to observe that the General tendency of your Management is to get the works hurried on without regard to the convenience of the public, the loss of the land occupiers or the advantage of the Company. You seldom provide the necessary accommodations before you begin to make the Bridges, and in the excavation you place your men in so many various places without either finishing as you proceed or making your fence wall & posts & railing – that the whole Country is laid open to damage. This grievance which has already caused so much trouble and expence is so very obvious & so much owing to your neglect that the Company will no longer suffer it.... The not fulfilling the promises you make & the want of attention in yourselves, your agents & workmen to the direction of the Company Agent, are evils there is great reason to complain of....[1]

The letter continues for some time in the same vein, and was followed by many others. The complaints went on and on – unfinished work, disregarding instructions, poor workmanship, puddling so badly made that it was impossible to repair and had to be done all over again, the contractors away at other workings (Pinkerton and Murray were also major contractors on the Leeds and Liverpool). In general, the contractors not only infuriated the Lancaster Company, they managed to infuriate everyone for miles around and caused endless rows with landowners. Astonishingly, the Company persevered with Messrs Pinkerton and Murray for two more years, though by the middle of 1795 the letters from the Committee were not so much despairing as almost hysterical:

[1] *Ibid.*, 14 December 1793.

> The clamour is raging that you are to have Barrow beck running into the Canal this day – in point blank contradiction to the Act of Parliament – for good Gods sake do not be so very crazy as to turn the said brook without having full liberty from Lord A. Hamilton.[1]

Rennie, at a good, safe distance from the troubles sent his word of consolation to the weary Gregson: 'I am not surprised at your being out of patience with P & M. I think it would be a hard business for Job himself.' By September 1795, the Company had had enough, and decided to settle up with Pinkerton and Murray: their work was measured by independent arbitrators, the final accounts settled, and the Lots readvertised. The final word came at the Company's General Meeting in 1796. Gregson described it to Rennie:

> Mr Pinkerton had an idea of creating some confusion & bustle amongst us & made a speech for that purpose – Dr Rigby answered him in the way he deserved & concluded with saying that what Mr Pinkerton had advanced were palpable falsehoods.
>
> The whole of the Proprietors of Lancaster knew Mr P was making false Statements from their knowledge of the work round Lancaster. No one seconded him . . . this I hope will be our closing scene with those worthy Contractors.[2]

The troubles were not all with Pinkerton and Murray, although they took the major share. Other contractors were almost as bad. Complaints followed the same pattern, which was repeated for many contractors and many canals. The contractors, as they were paid by the length of canal cut, pushed forward as quickly as possible, disregarding the quality of the work, the orders given by the engineer, and even whether the land had been acquired. Because they were so heavily committed in so many different parts of the country, the big contractors let their own work out to sub-contractors, which gave

[1] *Ibid.*, Archibald Millar to Murray, 10 June 1795.
[2] *Ibid.*, 11 January 1796.

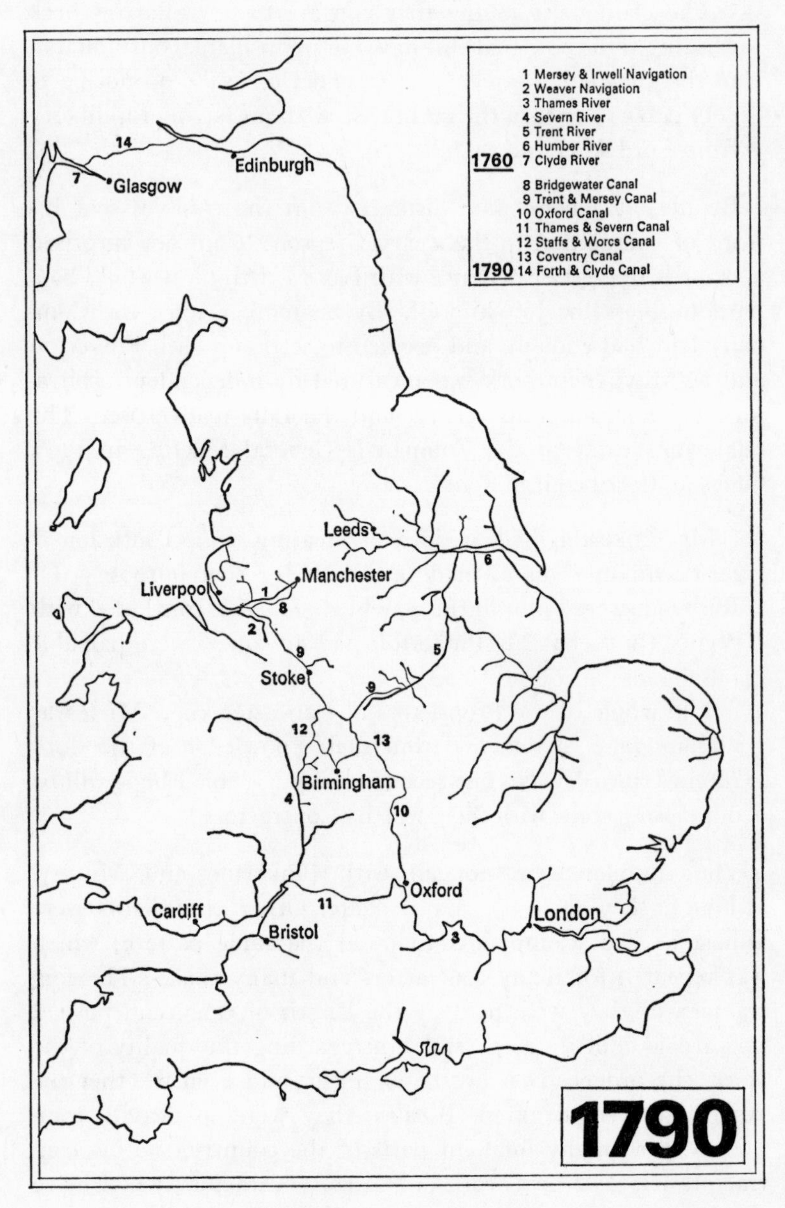
1 Mersey & Irwell Navigation
2 Weaver Navigation
3 Thames River
4 Severn River
5 Trent River
6 Humber River
1760 7 Clyde River
8 Bridgewater Canal
9 Trent & Mersey Canal
10 Oxford Canal
11 Thames & Severn Canal
12 Staffs & Worcs Canal
13 Coventry Canal
1790 14 Forth & Clyde Canal
Edinburgh
Glasgow
Leeds
Manchester
Liverpool
Stoke
Birmingham
Oxford
Cardiff
Bristol
London
1790

5 The waterways in 1790.

the engineer another opportunity to try to get the work under his own control:

> As it appears that Mr Pinkerton means to continue farming his contracted for Masonry (as Mr Shaw says) to Labourers or anyone – would it not be well for the Committee, to take the Second Contract off his hands and farm the works themselves. The Committee may depend on bringing better farmers on the works than Mr Pinkerton can. It is true they will give somewhat more to these Workmen, than what may be expected that he will – However these people will be more under proper Management, and in course better work may be expected.[1]

The engineer was, as usual, unsuccessful. The canal companies continued to let out the work and continued to argue with the contractors to whom they let it.

The contractors were also employers, and had their own problems with their own employees, usually about money, sometimes about not providing work for the men they brought to the site. As the contractors' main asset was the work force they controlled, it was essential that they always had enough men on hand to cope with the work. It was better, from the contractors' point of view, to have men standing around idle waiting for work than not to have enough. The men naturally took a different view. When John Murray kept his workers and subcontractors waiting for work, the men took their complaint to the Lancaster Committee and appealed for support. The Committee declined to interfere. As a general rule, the Company avoided getting involved in disputes between contractors and labourers unless they could see any particular advantage in doing so.

The main area of conflict between contractor and workers was pay – contractors not paying enough or, in some cases, not paying at all. It was a battle that was seldom won by the workers. The one period when the odds were on their side was harvest time, when they could always find alternative work on the

[1] *Mr Millar's Reports to the Lancaster Canal Committee,* 25 February 1794.

farms. Then the contractor might find himself losing his assets and having to face an irate Committee. John Plant, a contractor on the Warwick and Birmingham Canal, found himself in that position and received a curt note from the Clerk of Works: 'Notwithstanding I have furnished you with a considerable quantity of materials, for the purpose of forwarding the cutting at Hatton – you have not procured nor do I think you endeavour to procure men to work with them. . . . You may therefore expect to be discharged if the runs are not full on Monday next.'[1] For a few weeks the navvy held the better hand. Archibald Millar, during the time when he achieved his ambition of having workers employed directly by the Company, found himself faced by the same problem, and neither whip nor carrot worked:

> On Saturday last Mr Cartwright endeavoured all that he could to have all the Carpenters & Labourers at work, flattering them with Encouragement of Beer on the one side and threatening those who did not come should have no more Employment at the Aqueduct. The Scots men, Particularly the Carpenters, paid no respect.

He commented darkly:

> I hope and expect a little time will correct these combinations. The Harvest will not last always.[2]

The workers were well advised to get what they could when they could, for they could expect very little comfort at any other time. The one occasion when contractors and the Company could be guaranteed to find common cause was when they faced the workmen:

> Some of Cockran's men were here yesterday and troublesome for money. They wanted me to promise that I would see them paid, but this I refused to his cutters, but I promised the men who were digging the bed of the Calder, that I would

[1] *Warwick and Birmingham Canal Letter Book*, 30 August 1799.
[2] *Mr Millar's Reports*, 2 September 1794.

28 A fine collection of gongoozlers at the Marple aqueduct on the Peak Forest Canal.

29 The opening ceremony: scene on the Grand Junction at Paddington, 10 July 1801.

30 Two Canal Company seals which point the connection between the canals and industry: the Trent and Mersey seal shows a pottery with a conical kiln; the Ashby de la Zouch seal shows, behind the castle and the allegorical ladies, a canal passing a coal mine where a pumping engine and a horse whim are at work.

31 A beginning and an end: the Liverpool and Manchester Railway, opened in 1830, crosses the Bridgewater Canal, opened in 1761.

take care they should have their Money, provided they would return to the job and get it finished.[1]

The assurances given to the men on the Calder works was not due to any sudden flush of good will – the Company was prepared to make concessions where really urgent work was concerned. When there was no pressing urgency, then Company and contractor joined forces:

> Mr Stevens has some apprehension that a considerable part of his men intend turning out on Saturday (being his pay day) for an advance of wages. Should they do so, he is determined not to advance their pay, and he hopes the Contractors on your part of the Line will not employ any one who deserts from him. You will please to mention this to the Contractors on your part. If they give encouragement to Stevens' men it may be a real loss to the Concern by delaying the work here.[2]

The Company was prepared to back the contractors against the navvies, and the contractors were prepared to return the compliment. The result for the navvies was inevitable:

> Yesterday Morning the Companies Labourers hearing that Mr Stevens Men was paid 2/6 p day in consequence of which they all immediately left of their work and demanded 14/- p week and to work only 10 hours to the day: If they worked any more time to be paid 3d p hour. I hereby recommend to the Gentn of the Come to pay and discharge the whole of them tonight, and also Mr Stevens' Masons as Mr Millar and Myself have engaged with Mr Murray to send us 100 Men on Wednesday morning.[3]

In canal work everyone seemed to be preoccupied with money: the Company struggled to raise funds and then struggled to make a profit, the contractor did his best to squeeze the maximum profit from his contract, and the navvy tried to get the

[1] *Lancaster Canal Letter Book*. Letter from Gregson, 25 July 1798.
[2] *Ibid.*, 28 April 1796.
[3] William Cartwright to the Lancaster Committee, 22 April 1794.

semblance of a decent wage for his work. Canal companies and contractors often succeeded, the navvy seldom. As long as there was a surplus of workmen, the most likely reward for the navvy trying for higher pay was the sack. The law of supply and demand was ruthlessly applied.

In theory, the contractor was responsible not only for the navvies' pay but for their behaviour as well. He was supposed to exercise some sort of control over their activities. Samuel Gregson wrote to the contractors Vickers and Lewis to complain about their workmen:

> I am very sorry to be informed that some of your men have been behaving in riotous and very illegal manner at Chorley. Such conduct will bring disgrace upon the works, as well as upon yourselves, and every appearance ought to meet with an immediate and serious check. You will probably be called upon by the Justices in the neighbourhood to deliver up the persons who have been guilty, and I would advise you to show the utmost willingness in detecting and bringing them to Justice and be ready to do everything in your power which may be required to satisfy the publick for the Injury done in thus disturbing the Peace and assaulting the Publick Officers. This conduct in you will please the Committee and it is certainly in your Own Interest to keep your men in proper Subjection. I should advise you to discharge the whole who have been concerned if they will not make proper submission. Perhaps it would be right in you to take up some of the ring leaders yourselves and deliver them to the Justices.[1]

There is no evidence that the contractors took such responsibilities seriously. They were happy enough if the men worked when they were supposed to, and showed very little inclination to concern themselves with what the men did afterwards. Even Gregson's letter is not as stern as it appears at first glance. The recommendations for firm action are hedged about with conditions. It shows none of the direct 'dismiss the lot' approach that was shown when there was a threatened strike for more pay.

[1] *Lancaster Canal Letter Book*, 4 July 1795.

A few contractors found they made very little profit out of their canal work. Some were incompetent, others found difficulties in the works which they had not allowed for. Sometimes provision was made in the contract for exceptionally difficult conditions, and extra payment was authorized if 'any Rock or Quicksand, Bog Stone or Gravel which the Committee or their Engineer shall direct to be removed or hereafter mentioned be met with'.[1] When no allowance was made, or the allowance was not large enough, the contractor could always try appealing to the Committee or the engineer:

> In my report of the 24th June 1805 I stated that Thomas Wilkins had represented to the Sub Committee his loss of £800 on a Contract of Deep Cutting at Burbage. I am doubtful that for want of Capital he cannot carry it on with that Spirit he ought to do. If the Committee should think it right to advance him about the same sum to purchase Horses, Carts, & other Materials for carrying on this work, it will be of great Service to him.[2]

Some contractors who ran out of funds were scrupulous in trying to make a fair settlement with the company and with their own men. Others were less so. In the same month of July 1791 two contractors on the Leeds and Liverpool canal went bankrupt. One offered to have his work measured and to take that money and whatever he could get from the sale of tools and equipment to pay his men; the second absconded, not only leaving large debts behind, but taking the company's money with him – the company offered a reward and put advertisements in the local and London papers.[3] Many contractors when funds ran low simply got up and went, leaving whatever equipment they possessed and whatever debts they had acquired.

[1] Contract between William Montague and the Gloucester and Berkeley Canal Company, 7 October 1794.

[2] *Reports to the Kennet and Avon Management Committee.* Report by John Thomas, 24 June 1806.

[3] *Leeds Committee Minute Book, Leeds and Liverpool Canal Company,* July 1791.

Although there was an element of financial risk in being a canal contractor, there were more successes than failures – with the possibility of making a very great deal of money indeed. Even a medium-sized contractor, if he was careful, could make a handsome profit. The first impression given by a typical medium-size contractor's account book is invariably of a very small-scale operation. Thomas Sheasby, for example, was a contractor on the Coventry Canal between 1788 and 1789. The entries nearly all concern insignificant items of expenditure:

December 13, 1788 paid for 3 barrows and 1 Tressell – £1–8
gave men to drink at Sundry Times 3–6
August 8, 1789 Ale for men whilst puddling 11–6

From the entries, he appears to have employed, on average, no more than twenty to thirty labourers – but his contracts were worth a total of £17,040.

Thomas Sheasby could expect a decent profit from his business. The big contractor would expect an enormous one. Some Companies offered bonuses to the big contractors. Jonathan Woodhouse had a tunnelling contract with the Huddersfield Canal Company which gave him a £400 bonus for every month saved – but carried a £200 forfeit for every month lost.[1] The competition for such lucrative profits was intense. The contractors relied to some extent on getting advance information and a personal recommendation from the engineer. Pinkerton received advance notice of the Lancaster lettings from Rennie – an act which the engineer must have soon started regretting. The contractors were not above trying to help the process on its way:

> Whereas . . . Mr Greenway was ordered to prepare the Draft of a Contract between the Engineer of the said Company and Robert Pinkerton for taking the Earth out of the Hills at Rowington and Shrewley on the line of the said Canal and embanking the Valley between the said Hills . . . and whereas it has been made to appear to this Committee that

[1] *Huddersfield Canal Company Committee Book,* 23 December 1802.

> the said Robert Pinkerton endeavoured by undue means to injure this Company by making improper offers to their Engineer. It is therefore Resolved that the said Robert Pinkerton shall not be employed.[1]

From a rather broad hint in a letter from Archibald Millar to Pinkerton and Murray, this was not an isolated instance:

> All the good you can do me *as an individual* [Millar's italics] will never widen my conscience nor cause me to swerve from what I think is right.[2]

Apart from bribery, the contractors were certainly not above helping themselves to any useful materials they might happen to 'find'. Stevens, the contractor on the Lune aqueduct, took a load of sand from Pinkerton's lot, and, when the latter found out, began to take sand from the near-by roadworks instead. Contractors were just as happy to cheat each other as they were to cheat the Company. Millar, in trying to settle another dispute between the same two contractors, commented, with some justification: 'Gentlemen, I very plainly forsee Greed in both sides.'

The contract system had its obvious disadvantages. As prices were fixed before work began, there was a temptation to skimp work in order to ensure the maximum profit, and the canal engineer needed to keep a keen eye open. The system also provided opportunities for the more dishonest contractors to steal, cheat, and bribe their way to a profit. The contractors were hard employers, but it was an age of hard employers, and it is doubtful whether the navvy would have been any better off under a different system. On the credit side, it was probably the most efficient system that could be devised for building a network of canals, each section of which was built by a different company. Under the contract system, men could easily be moved about the country to where they were most needed and, most importantly, as gangs of experienced workers were

[1] *Warwick and Birmingham Canal Minute Book,* 31 July 1794.
[2] *Lancaster Canal Letter Book,* 4 April 1795.

moved so their experience could spread. No canal works had to worry about suffering from a complete absence of experience – even if there was no local experience to draw on, they could go farther afield and bring in experience from outside the district.

15

MEN AT WORK

Here smooth canals across th' extended plain
Stretch their long arms to join the distant main.
The sons of toil, with many a weary stroke,
Scoop the hard bosom of the solid rock;
Resistless thro' the stiff opposing clay
With steady patience work their gradual way;
Compel the genius of th' unwilling flood
Thro' the brown horrors of the aged wood:
'Cross the lone wastes the silver urn they pour,
And cheer the barren heath or sullen moor.

'One of our first female poets', quoted in J. Phillips, *A General History of Inland Navigations.*

If any of the 'sons of toil' had ever read the lady's poem – an unlikely occurrence – they might have agreed about the weary strokes and the stiff, opposing clay – but the idea of the steady, patient navvy quietly working away at it would have seemed quite ludicrous. The navvy was far from quiet or patient, and his work was unpleasant and often dangerous.

Canal cutting was unrelieved hard labour. In the early years, there was nothing complicated about the work: men were simply set the task of digging a deep, wide trench through the land. They piled the earth into barrows, which were then wheeled away and emptied. It was hard work, but required no special skills – the only advantage that the experienced navvy had lay in the years of practice which enabled him to work faster and longer than newcomers to the diggings. No more than that was required of the diggers as long as the canals kept to an even course, winding their way round any hills or other obstacles. But canal engineers became more adventurous. Instead of going round every obstacle, they began to cut straight through them. The making of a deep cutting through a hill called for a new technique, and the barrow runs were introduced into the canal scene.

In a deep cutting, it was impossible for a man simply to throw earth into a barrow and wheel it up to the surface: the sides of the cut were too high and steep. So, first of all, planks had to be laid up the side of the cutting to take the barrows. Then the barrows had to be hauled up to the surface, usually by a horse. When the barrow was full and ready to be taken up, the horse would set off at a steady pace and the barrow would be pulled slowly up the run. The job of the man on the barrow run was to walk up the steeply-inclined plank, which was probably coated with slippery wet clay, while keeping a loaded barrow steady in front of him. If the horse suddenly changed pace, or the man slipped, then down went the barrow-load of earth back into the workings and, in most cases, the man went with it. Not a serious accident compared with some that could happen to a man on canal work, but there are pleasanter ways of spending a working day than sliding down a bank of clay with a barrow load of earth on top of you. The men on the runs worked all day trudging up and down the slippery planks, balancing the heavy barrows in front of them, and, with runs of fifty yards or more, such accidents must have been commonplace.

By the end of the eighteenth century, some of this work had been taken over by machines. A report in *The Times* of August 1796 described a machine being used on the Gloucester and Berkeley Canal, which appeared to work on the conveyor-belt principle – the barrows were hauled in a continuous chain, and it was claimed that, with only two men to work it, 1,400 barrow loads could be shifted on a forty-foot run in twelve hours. Some of the work on the Caledonian was done by steam dredgers – but, even there, on the most highly mechanized of all canal workings, the major part of the cutting was done by men and horses on long barrow runs. The effect of machinery was minimal.

The earth that came out of the deep cuttings was often taken away to be used for building up an embankment in the adjoining valley. In the first stages of the work, the earth was loaded into carts to be taken from the hill site to the site of the embankment. This was a massive operation, which involved many different resources. The work on the embankment that brought the Lancaster canal to the aqueduct at Lune was typical of many others all over the country. Full reports of the progress of the embankment were sent to the engineer Rennie. At first, all the earth came from the cutting at Ashton Hill where 'six stout sets of contractors' were at work with 'another set at the lower end'. The earth was then taken and, according to the reporter's estimate, 'near 50 Horses and Carts will be employed raising the Lune Embankment from the adjoining fields'.[1] Earth for the embankment came from other parts of the works as well: 'We have entered the field on the lower side of the Bulk aqueduct and are carting the whole of the upper part into the Lune embankment, which will yield about 40,000 yards of Earth.'[2] Once part of the canal could be filled, the work of shifting earth from cutting to embankment could be done by 'boating' instead of carting:

> Our Boats from Ashton hill to Lune Valley make Two [journeys] p day now to the Wyer Embankment carrying 12 yds each, will finish in 18 weeks, 5 working days p week. Two Horses, Two Steerers, Two boys and forty-three fillers and emptiers will with ease do the work in the above time.[3]

In other works, rails were laid and the earth was shifted in strings of wagons running on the temporary tramway.

The embankments are often forgotten when assessing the achievements of the canal builders, but they represented a major part of their work. In its own way, the huge embankment

[1] *Lancaster Canal Letter Book*, 1 November 1795.
[2] *Ibid.*, 18 November 1795.
[3] *Ibid.*, 8 October 1796.

that brings the Ellesmere canal to Pontcysyllte is as impressive a monument to the canal workers as the great aqueduct itself.

It was not just soil that had to be shifted out of the hills – the navvies also had to move rocks. In a few cases, as in the Rochdale excavations, entire cuttings were made by blasting through solid rock. The stone was used, as was the soil, in constructing other parts of the canal.

Once the channel was dug out, it had to be puddled to make it watertight. Puddling was one of the most exhausting parts of the navvy's work. First, suitable clay for the puddle had to be found, dug out and brought to the section where it was needed. This in itself could be a major operation:

> The makings of Linings are tedious owing to the distance in many places from which Clay is to be brought. We therefore propose to make a Rail Road between the Deep Cutting at Devizes & Horton (Where Gregor is now boating) and put on another Contractor to make up abt a mile of Lining.[1]

The clay, when it arrived, was mixed with water and the mixture applied in successive layers to the bottom and sides of the channel. The most common method of pressing in the semi-fluid puddle was to stamp it in with the feet. It requires little imagination to see how tiring a day spent tramping up and down in wet clay would be.

Cutting and puddling finished, the workers levelled off the towpath, put up fences and let the water in. The main part of a canal navvy's time was taken up with this type of heavy labouring work – digging out the canal, working the barrow runs, filling and emptying carts and boats, puddling. It was the unspectacular but essential part of canal building, and, in the frantic years of the 1790s and into the early 1800s, there were thousands of men all over the country working in this same way. It was not, however, the only part of canal construction. Canals also involved building bridges and aqueducts and tunnels.

. . . . .

[1] *Reports to the Kennet and Avon Management Committee,* letter from John Rennie, 25 September 1804.

Aqueducts are probably the most attractive and interesting features of canals. The biggest were given the full treatment in accordance with their status – ornamented with battlements or embellished with classical columns – for they were rightly regarded as the engineering triumphs of the eighteenth century. Building these beautiful structures, however, was as much a matter of hard graft as cutting and puddling was.

The Lune aqueduct, like all the big aqueducts, took the canal across a river. This meant that the foundations had to go down into the river bed. The first stage was to build a coffer dam to keep the water out of the workings, and allow the navvies to begin the arduous job of pile driving for the foundations. Work on the Lune foundations began in January 1794, under the direction of Archibald Millar. The carpenters' shop and saw pit were prepared and the steam engine that was to pump water from the workings was brought in. Exley, the superintendent, moved into his temporary home at the site, and a blacksmith and a carpenter joined him. The carpenter, 'a person of integrity', was also to act as foreman at a suitably elevated salary: '18 or 20/- a week Certain would be a very pretty inducement to a good working man.'[1] By March, they were beginning work in earnest, but things did not proceed with perfect smoothness. There were strikes[2] and a number of hold-ups when the river flooded, overflowing the coffer dams and, in one case, carrying planks, piles and barrows away down-stream. Even when things were going well, the men who worked at the pile driving had to do so in the most atrocious conditions, struggling in the mud behind the coffers in a pit which the pumps could never keep completely dry. The work of pile driving was hard and exhausting: as Millar reported to the Committee, 'it may be proper to give some small Sum dayly to the Pile Drivers in Drink'. The job of cutting and sawing the massive piles was very little easier. The pile-driving engines, which had been constructed at the site, were crude, man-powered devices, and it is no surprise to find this entry for 6 May:

[1] *Mr Millar's Reports to the Lancaster Canal Committee*, 28 January 1794.
[2] Described in the last chapter.

> A labourer at the Lune Aqueduct had the misfortune yesterday to have three of his Fingers on the right hand taken off by the Piling Ram falling upon them. I should recommend him to the attention of the Committee to give him a small sum to assist him in his present situation.

The work on the aqueduct foundations, and the work to bring the canal up to the site, still continued at a reasonable pace into the Summer of 1794. By June, work on the coffer dam for the second pier had begun, and there were over 150 men at work – '127 labourers, 22 Carpenters and 14 Sawyers'.[1] The canal cutting and embanking was being pushed along as fast as possible with double shifts of men, working day and night. The following month, work on the second pier was so far advanced that there were eighty labourers at work in pile driving alone, and the Company was again doing its bit by way of encouragement: 'a little Beer will be necessary given to the Workmen to stimulate exertions, that article having been stopt for a time'.

As the foundations were completed, so the next stage could be started. The masons came in and the scaffolding started to rise in readiness for the work on the arches. The area round the aqueduct was alive with activity. Hundreds of navvies were at work in cutting, banking and pile driving. The steam engine was pumping night and day. The saw pits and carpenters shop were busy with the manufacture of piles and scaffolding. The masons were dressing stones for the piers. Carts passed continuously, bringing supplies of stone and timber. The navigable part of the canal was being used for boating earth to the embankment. Altogether, the Lune had become the focal point for activity on the Lancaster canal. During that period there were almost a thousand men at work in the whole concern and over two hundred horses and carts were being used to carry materials. It was a scene that could be repeated many times in many places: and the Lune was only one of a half-dozen or so major aqueducts being built at the same time.

Big aqueducts, like the Lune, represent the most obvious part

[1] *Ibid.*, 24 June 1794.

of large-scale constructions. When it was finished in 1796, it had cost the Company £48,000. Hundreds of workers had been involved in its construction, and material had been brought from as far away as Italy, the source of the 'pozzolana' – the water-proof mortar used in the foundations. The big aqueducts, though, were the exception. All over the country, small aqueducts were being built, most of which go unremarked. How many of the thousands of motorists who use the Holyhead Road each day realize that one of the many bridges under which the road passes is no bridge at all, but an aqueduct carrying the Birmingham and Liverpool Canal over their heads?

Of all the canal works, those that took the greatest toll in terms of effort and hard work – and injuries and deaths – were the tunnels. From the very first, the canal tunnel was an engineer's nightmare. Brindley achieved the construction of his 'impossible' aqueduct at Barton with little trouble, but he never lived to see the end of the eleven years that were spent on the old Harecastle tunnel. Yet, as a piece of engineering, the Harecastle was crude, unfinished, a rough hole through the hillside. But its 2,900 yards all had to be cut through the most difficult ground with only the most primitive of techniques.

The technology of tunnelling came from the mines, and the men who did the skilled work in canal tunnelling mostly came from the same source. The basic plan of work was the same for most tunnels. At first, shafts were sunk down from the hill to the depth of the tunnel and trial borings were taken. The men began work at the entrances, and as they progressed into the hill, the shafts were used both for taking waste and rubble out of the works and for lowering and raising the workers themselves. Horse gins were used for this job: they were simple machines – a horse walking a circular path was harnessed to an arm of the gin; as the horse circled it turned a drum around which a rope was wound, and this in turn raised and lowered the bucket in the shaft. The rock was blasted out with gunpowder, but the main part of the work was still done by men with picks and shovels. Steam engines were usually used for pumping to keep the works dry, and in later tunnels rails were often laid to ease the job of taking rubble out and bricks for lining in, when the excavation was complete.

The Reverend Shaw visited the works at Sapperton tunnel on the Thames and Severn Canal in 1788:

> We now approached the great Tunnel, which forms part of the communication between the Severn and the Thames; on each side of the road it extends rather more than a mile; one end penetrates the hill at the village of Sapperton, the other comes out in Heywood; we turned on our left to visit the former, and saw the shafts busy in several places, at the distance of about 230 yards from each other; by this means they wind up the materials from the cavity and expedite the work. The earth is principally a hard blue marle, and in some places quite a rock which they blow up with gunpowder; the depth of these pits are upon an average eighty yards from the surface. The first contractor receives £7 per yard from the company, and the labourers rent at the rate of about £5 per yard, finding candles, gunpowder &c. The workers are in eight gangs, and continue eight hours at a time, day and night.[1]

He did not go into the workings at Sapperton, but he visited various mines where conditions were very similar to those in the canal tunnels:

> Thus far in the mountain, with the aid of lights, 'tis easy enough of access; but such a horrid gloom, such rattling of wagons, noise of workmen boring the rocks under your feet, such explosions in blasting, and such a dreadful gulph to descend, present a scene of terror that few people, who are not versed in mining, care to pass through. . . . On the passage down, the constant blasting of the rocks, louder than the loudest thunder, seems to roll and shake the whole body of the mountain . . . the glimmering light of candles, and suffocating smell of sulphur and gunpowder, all conspire to

[1] Rev. S. Shaw, *op. cit.*

increase surprise and heighten apprehension . . . at the same time figure to yourself the sooty complexions of the miners, their labour, and mode of living, and you may truly fancy yourself in another world.

The self-induced and pleasurable horror that Shaw describes is something of a stock reaction from the eighteenth-century gentleman, but the conditions in the narrow tunnel workings were truly terrible. Even in the 1830s, with sixty years of tunnelling experience in canal works to draw on, the railway tunnellers worked under the same conditions:

> On some occasions the miners refused to work. . . . Nor is this surprising, considering the nature of the operation: boring their way almost in the dark, with the water streaming round them, and uncertain whether the props and stays would bear the pressure from above till the arch-work should be completed. Those who visit the Tunnel in its present state . . . will not easily picture to themselves the original dark and dangerous cavern, with the roof and side supported by shores, while the miners pursued their arduous task by the light of a few candles, whose feeble glimmer glancing on the water which ran down the sides, or which spread out in a sheet below, was barely sufficient to show the dreariness of the place.[1]

Work on the great tunnels went on night and day, and, indeed, in those gloomy holes there was very little difference between the two. And they certainly were great – Sapperton tunnel was 3,808 yards long and was only surpassed by Standedge tunnel on the Huddersfield Canal which stretches for an astonishing 3 miles, 176 yards. What extraordinary self-confidence those engineers showed who dared even to contemplate making such a tunnel which, at its deepest, is seven hundred feet below the surface, when they had so little experience and such crude techniques and machines. Altogether the canal

[1] Henry Booth, *An Account of the Liverpool and Manchester Railway* (1830).

engineers built twenty tunnels that were more than half a mile long, and many more shorter ones.

It was in tunnelling that the worst accidents happened: the falls of earth, trapping and suffocating a man; the collapse of props or scaffolding; the damage from blasting; the inevitable accidents that happened when men took the precarious ride up and down the shafts, travelling in buckets suspended by ropes. The Company records seldom mention the accidents that happened in the works – but parish registers show a number of deaths to 'strangers' that coincide with the period of canal construction in the district. Occasionally, there is mention of compensation – four workmen injured when scaffolding collapsed inside the Foulridge tunnel on the Leeds and Liverpool had the surgeon's bill paid for them by the Company and received compensation (two received a guinea and two had half a guinea each). But canal companies and contractors were not particularly noted for their benevolence towards the men who were injured and maimed in the workings. When they did pay up, it was hardly with a good grace:

> The inclosed paper the Committee refer to you for your Opinion and Observation thereon – They think the charge of Six Guineas for the attendance of the Surgeon is too much – about one half of that sum may be a reasonable claim. It is also their opinion that Messrs Vickers & Lewis ought to pay this charge, as the man received his misfortune in their employ – however if the Doctor comes to reasonable terms they have no objection to pay One half (privately) & Messrs V & L the other half. You must manage this so as not to make it Public, or every man who receives any hurt upon the whole Line will become a claimant upon the Company.[1]

The Huddersfield Canal Company was in charge of the biggest of the tunnelling projects and, although they kept no records of the accidents that happened in the tunnel, they must have been very frequent, on the evidence of this entry in the minute book: 'Whereas several of the Workmen of this Company have been

[1] *Lancaster Canal Letter Book*, Gregson to Eastburn, 10 February 1796.

much hurt and bruised in the Company's Works, it is thought that Mr Rooth should subscribe to the Manchester Infirmary.'

If the smaller aqueducts remain unnoticed, the bridges and locks are such an accepted part of the canal scene that it is easy to forget that they all had to be built – someone's labour had to be used. The construction of a lock involved a mixture of hard work and specialized skill. The excavation of the chamber certainly involved the first item in large measure. Even the normal locks required a vast amount of earth to be shifted and when it came to the big locks, such as the sea locks on the Caledonian Canal, the digging became as arduous a job as any to be found in the whole workings.[1] The masonry work of the lock chambers involved craftsmanship of the highest order in the dressing and placing of stone blocks, and many an old lock gate remains today to testify, by its combination of massive timberwork and perfect balance, to the skill of the canal carpenters. In the course of canal construction, the Company would normally have to supply two or three bridges for every mile of waterway – put all those together and you have a figure for the whole system of something like ten thousand canal bridges. That represents a great deal of labour and a further call on the resources of bricklayers and masons.

Separate from the canals, but a vital part of the system, are the reservoirs. The Leeds and Liverpool Canal, for example, had seven reservoirs with a total capacity of almost 1,200 million gallons, and each reservoir required its own separate feeder channel to take the water down into the canal. This represented more than a major building operation, it was work that called for a high degree of civil engineering skill: more indeed than it always received. There are many reports of the bursting of reservoirs: the reservoir on the Crinan Canal burst in January 1811, which 'entirely interrupted the navigation of this canal for several months' and when the Driggle Reservoir burst on the Huddersfield Canal, six lives were lost.[1]

[1] See below, p. 122.

[2] *The Times*, 9 November 1811 and 11 December 1810.

Canal work gave rise to a host of subsidiary activities, for the canal concern had to be a self-sufficient unit. As far as possible machinery was constructed *in situ*; even a simple device like a crane required a lot of work: 'Pd John Parrish Expences for himself & Horse 3 Days going to Coalbrook Dale to procure Iron-Work for a Crane & Stops.'[1] Quite complicated machines might have to be 'home-made'; there is a tradition at the pumping station at Claverton on the Kennet and Avon canal that the balance beams for the pumps were cast at the site. This would certainly be in keeping with the canal builders' practice of keeping as much as possible of the work under their own eyes. All the material for building, for example, would normally be provided by the company. Stone for bridges, wharves, lock cottages and other buildings would normally come from the works themselves or from the company's own quarries. This, in turn, involved the company in hiring quarry men or taking on extra contractors, and also involved them in building roads, however crude, from the quarries to the works. Bricks were also used in enormous quantities, and they too were made on the site. Rennie had some problems in acquiring bricks for the Kennet and Avon:

> I had intended to contract with Christopher Shew who was recommended from Bristol for another 1,500,000 and he had been digging clay for the purpose, but I found him so earnest to draw Money, that I hesitated, and at the last Payment, he went off & though he is not in Debt to the Company, I am sorry to say he has left his Workmen and Tradesmen in Debt.[2]

This was soon sorted out, however, and the work went ahead: 'We have burnt some bricks at the foot of Devizes Hill . . . we have erected 3 Kilns in the best field & shall now have a regular succession from this Time & good hard bricks.'[3]

So the lists of workers and tradesmen connected with the

[1] *Coventry Canal Journal*, 27 May 1770.

[2] *Engineers' Reports to the Kennet and Avon Management Committee*, 20 December 1802.

[3] *Ibid.*, 27 January 1803.

canal grew: navvies and miners, masons and brick-layers, engine-men, carpenters, blacksmiths and so on to cover a vast range of different activities.

The link between the different parts of the canal and the associated works were the carters, the men who shifted the bulk of the essential materials. Although the carters performed an essential role in the complicated system of construction, they also did more than any other group to antagonize the neighbourhood through which the canal passed. The effect of having upwards on two hundred heavily-laden carts added to the local traffic of already inadequate roads was bound to be disastrous. It was made even more so by the carters' indifference to the effect they were creating – for as long as they were paid by the amount of work they got through they were not going to spend too much time worrying about local susceptibilities. As far as they were concerned, it was all very well for the Company to complain about practices such as using narrow-wheeled carts for heavy loads – narrow-wheeled carts might damage local roads, but they were quicker and time could be measured in cash. The same complaints about carters were constantly reiterated; carters often had control – nominally – over more than one cart at a time, and they were not too fussy about the horses and carts at the back of the procession causing damage or blocking the roads so long as they all arrived at their destination. Carts damaged roads and bridges; and there were endless complaints from farmers who found their stock spread all over the countryside because the carters had left gates open. In general they were a nuisance. But they were necessary.

Among all this varied activity, the bulk of the work still fell to the ordinary navvy. Not that the navvy would have thought of himself as 'ordinary'. He almost certainly regarded himself as a prince among workers. He could work harder and longer than any other worker, a fact which he had constantly demonstrated to him when new, inexperienced labourers came to the diggings. He could also live harder and fight harder, a fact demonstrated in battles with other workers, or even in prize fights:

> Tuesday morning, about nine o'clock, a pitched battle was fought behind the Ben Johnson's Head, Stepney-fields, between Samuel Elseworth, a butcher, in Shadwell, and Ned, a navigator, for two guineas a side, which lasted one hour and seven minutes; and after forty-six rounds, Elseworth gave in, being unable to move . . . he died the next morning.[1]

The canal navvy's boast might have been justified, but how did his pay and working conditions compare with those of other workers of the time? Did the prince of workers get a suitably 'royal' wage?

Wages for the first canal workers were at the same sort of level as those for other workers. The navvy, as such, could hardly be said to exist – he had not yet acquired any special skills, so there was no reason for him to get special rates. An entry in the Oxford Canal Committee Book for January 1770 reads: 'Inasmuch as there are 700 men now employed in the works and their number will be still increasing daily so as to require that a sum of £50 at least p Day sho^d^ be set apart.' Assuming that they were working on the theory that numbers could be going up to around the thousand mark, that would work out at an average of about a shilling a day, certainly not as much as 1/6. Rates to contractors on the Coventry Canal at the same time were as low as 2½d a yard for easy cutting, only rising to 4d for the deepest cutting. If the contractors were to make any profit at all – and they expected, in fact, to make a fat one – labourers must have been paid at a considerably lower rate. This would support the idea that workers' earnings would not have been much above a shilling a day.

Over the next twenty years, canal workers' pay practically doubled. On the Lancaster canal 'from 220 to 230' men were reported to be at work near the hill at Corforth in October 1793 at 2/6 a day. This was a comparatively high figure, for wage rates did fluctuate considerably. There were no subtle bargains struck at the canal works: rates were decided by applying the laws of supply and demand in their crudest form:

[1] *The Times*, 15 February 1805.

> As the soldiers will be disbanded on Saturday I propose trying if we cannot get about 20 of them to carry up the cut. . . . If we cannot get any soldiers I propose that Wilkinson & Caton should each raise a gang & go up with them – to do this they must give extra wages for a time.[1]

If cheap labour was to be had, rates went down; if it was scarce the rates went up again. As a general rule, the rates stayed somewhere between a low of two shillings and a high of 2/6 during the 1790s.

The canal workers usually got rather more money than other workers in the surrounding countryside. Telford and Jessop explained why this had to be so in a report on the plan for work on the Caledonian:

> As canal work is so very laborious, they must give such Wages . . . as will be the means of procuring and calling forth the utmost exertions of able Workmen; so that although the Wages paid by the Contractors may be higher than for common Workmen in the adjoining country, yet when compared with the quantity of Work performed, it is by much the cheaper labour.[2]

There were good reasons for paying the canal labourer more than other labourers, but the effect was bound to be felt outside the canal works themselves. The higher rates of pay tempted men to take up the nomadic navvy's life in preference to staying on the land, so that farmers were forced either to put up their own rates or go short of labour. Lincolnshire farmers petitioned Parliament in opposition to the Bilston Canal Act in 1783: 'a great scarcity of labourers prevails during the time of harvest, and has, within a short period, very perceptibly encreased insomuch that wages have, in that neighbourhood, risen to upwards of seven shillings a day'.[3] The figure of seven shillings is certainly

[1] *Lancaster Canal Letter Book*, 22 April 1802.
[2] *Second Report of the Committee for Making and Maintaining the Caledonian Canal*, 17 May 1805.
[3] *Journal of the House of Commons*, 27 March 1783.

exaggerated and, in any case, the farmers soon found a way to come to terms with the situation. For most of the year, the farmers watched their men go off to try the more lucrative canal work, but at harvest time they put up their rates for the season, and the canal engineers had to take their turn watching their navvies leave for what had become, temporarily, the more lucrative work of harvesting.

But navvies were not the only canal workers. The skilled tradesmen – carpenters, bricklayers, and masons – and specialists, such as blasters, all got more money than the ordinary cutters. In 1794, rates on the Lancaster were quoted as 'Masons at a Masters price of say 3/-, quarriers at 2/6 and labourers 2/2'. The craftsmen also sometimes had the advantage of being employed on a regular basis: in 1768, when labourers were getting about six shillings a week, a full-time carpenter on the Coventry Canal was taken on at eleven shillings a week. There were also some lower down the scale than the navvy, particularly the boys who were employed to lead the horses at the gins and barrow runs. In 1824 they were being paid 1/6 a day for their work on the new Harecastle tunnel, which compares with a basic rate for labourers of 3/4 a day.

It is difficult to assess whether canal workers were 'well' paid for their work. By modern standards, of course, their pay was atrocious, but relating pay to cost of living is notoriously difficult for this period. It is, however, possible to see how the pay and working conditions of the canal worker compared with those of workers in other industries.

Canal workers were never the worst paid of the workers of the eighteenth century. They were better off than the farm worker and they were not the worst paid group in the industrial sphere either, that doubtful honour probably belonging to the framework knitters of the Midlands, whose position was so desperate that they twice petitioned Parliament for help, in 1778 and 1779. A workman claimed that to make a dozen pairs of stockings a week he literally had to work from dawn to dusk. For this he was paid seven shillings a week, out of which he had to meet his own expenses, including the hire of the knitting frame. He was lucky if he finished the week with five shillings. The poverty of the framework knitters was notorious, but it makes the nine shillings a week that the navvy could earn at that time look like quite a large amount, especially as the knitter was a well-trained and skilled workman.

The knitters were, perhaps, something of an exception: the most obvious comparison is with the archetypal figure of the industrial revolution – the cotton worker of Lancashire and Derbyshire. The same travellers who went to gaze at the new wonders of the canal age, went too to look at that other new interesting specimen, the factory worker:

> Turning back and passed a cotton manufactory; the people all coming out to dinner, for it was already one o'clock. From the glance I had of their appearance, the observations I made were these: They were pale, and their hats were covered with shreds of cotton. Exclusive of want of exercise, the general bane of all manufactures, the light particles of the cotton must be inhaled with their breath, and occasion pulmonary affections. Owners of factories should consider this; for it is not the extended view alone of benevolence (a view they little regard) but the contracted hope of self-interest, that ought to have some influence on them. Let every such person, then, order his work-people to bathe every morning, and let them have a piece of play-ground for them, wherein some athletic and innocent exercise might be enjoyed for an hour or two, each day. In cotton-works, let each of them drink much water; and to crown the whole establish a Sunday-school, where they might be instructed orally, without being taught to read. This I deem a necessary precaution, as they would have all the advantages of improvement of mind and morals, without their common banes – low political club-rooms, with their idleness, their liquors, and neglect of families.
>
> But how, say the manufacturers, is self-interest to be advanced by this? I answer, enter into a resolution not to give the present exhorbitant wages of 2s to the meanest hand, and yet keep them so far above the wages of agriculture, as to

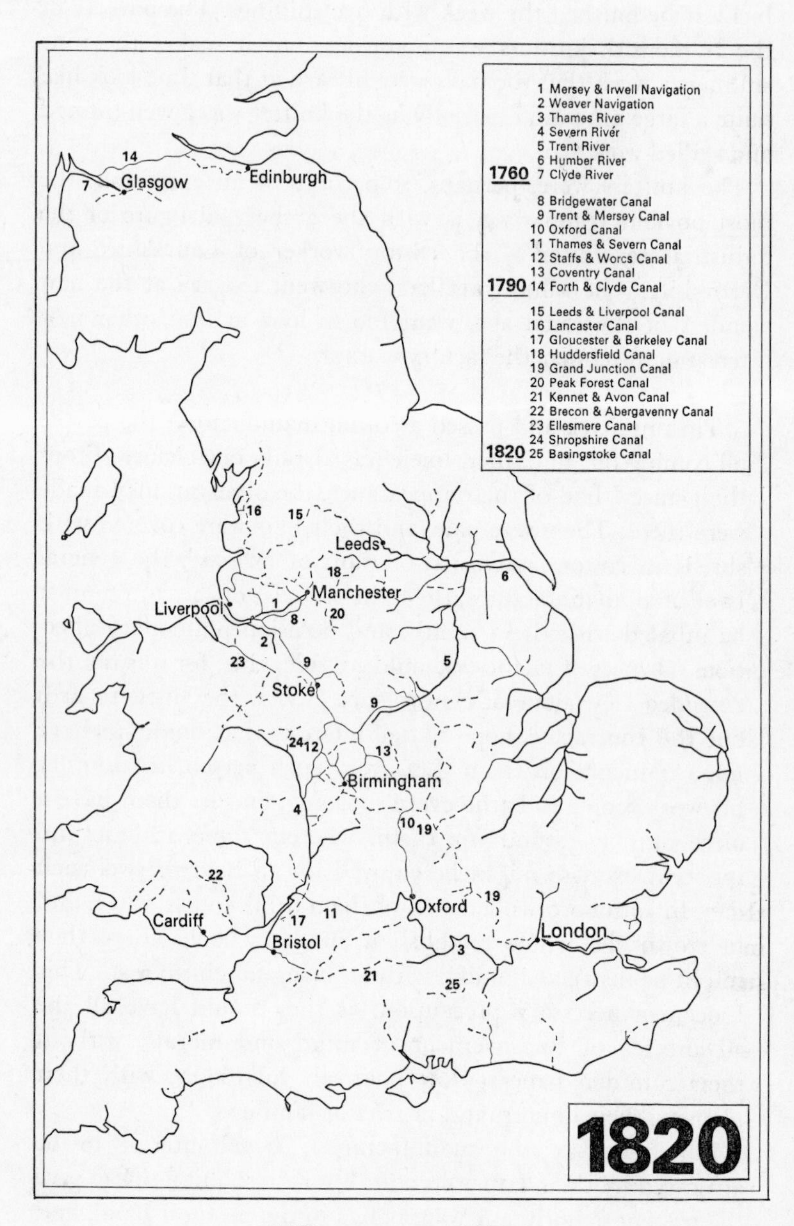

6 The waterways in 1820.

excite temptation. Where 18d is the day's receipt of a labourer, 2d makes a wonderful odds. At present how does the matter stand? Why thus; the manufacturer gets 4s in two days, and this enables him to ruin his health by drunkenness, and his mind by idleness on the third.

His money spent and his mind enfeebled, he returns unwillingly again to work, discontented, and cursing all laws, human and divine, which have so arranged matters that yon stately house, and the gilded coach in which its owner rides, should belong to what the Corresponding Society, the illuminati and illuminates of this country, have deluded him into the idea, is an individual, with no better claim to it than himself.[1]

The author of these fine, ennobling sentiments obviously considered himself far more benevolent than the factory owner – one needs to know little beyond that to envisage what the attitudes of the factory owners must have been. Taking two shillings a day as an 'exhorbitant' wage for the factory worker, then the navvy still comes out of the reckoning as comparatively well paid: when set up, that is, against the other labourers in other industries. As Grant says, when you are working that close to the starvation line, twopence a day makes a wonderful difference to a man. This was why men were prepared to abandon their homes to go to work on a distant canal. The canal navvy was also spared the worst of the degrading conditions that attached to the life of the factory worker. Perhaps too it was worth working in the wilds if it meant that you were spared the moralizing of such pious commentators. And the navvy had one great advantage over the other workers – he could get up and move on.

When work on the canal was ended, the navvy had his part in the official celebrations: he and his workmates marched in procession, cockades in their hats, shovels and picks over their shoulders. When the Wey and Arun Canal was opened in 1816

[1] J. Grant, *Journal of a Three Weeks Tour in 1797 Through Derbyshire to the Lakes.*

the 220 navvies sat down to eat their way through a roast ox and to drink their way through two hundred gallons of strong ale. The next day they were off again for the next job on the next canal, leaving behind them yet another part of the waterway system open for traffic.

16

GONGOOZLING
A CELEBRATION OF CANALS

Gongoozler: an idle and inquisitive person who stands staring for prolonged periods at anything out of the common.

H. R. De Salis, *A Handbook of Inland Navigation* (1901).

Although canal building is the subject of this book, it is only the beginning of the canal story. It seems a pity to end without at least a brief look at the end product of so much effort expended by so many people, and so this chapter takes a brief look at the canals through the eyes of a variety of gongoozlers. The curious spectators of the eighteenth and nineteenth century were little different from those of our own time: where a group of men were found at work, another group could always be found to stand and watch them. Even before the canals were finished there was no shortage of spectators, though they were not encouraged by the Company:

> The Lancaster Canal Committee request the Public in general, that their curiosity may lead them to trespass as little as possible on private property: Breaking or pulling down Gates & Hedges are offences at all times without excuse, and if the offenders are discovered they will be prosecuted as the Law directs.[1]

[1] *Lancaster Canal Committee Minute Book*, 15 January 1793: order for an advertisement to be printed and distributed.

But, when work was finished and the Grand Opening Day finally arrived, then anyone and everyone could come and stare. The Very Important got to ride in the boats, while the lesser beings had to make do with watching from the towpath. Everyone had a splendid time: the military bands played suitable patriotic and stirring airs; the proprietors made suitably pompous speeches; poets recited atrocious but lively verses, and the canal workers set about the serious business of getting drunk at the company's expense. The weather was, of course, fine; the scenery, inevitably, magnificent. Even profits were temporarily forgotten – at least, so the commentators claimed. Here is the whole magnificent scene described for *The Times*:

> On Monday last, the navigation of this Canal, from the Thames to the town of Croydon was opened. The proprietors . . . met at Sydenham and there embarked on one of the company's barges, which was handsomely decorated with flags, &c. At the moment of this barge's moving forward an excellent band played 'God save the King', and a salute of 21 guns was fired. The proprietors' barge then advanced, followed by a great many barges, loaded some of them with coals, others with stone, corn &c, &c. . . .
>
> The gay fleet of barges entered Penge Forest. The Canal passes through this forest in a part of it so elevated, that it affords the most extensive prospects, comprehending Beckenham, and several beautiful scattered villages and seats. . . . The proprietors found their calculations of profit irresistably interrupted, by these rich prospects breaking upon them from time to time, by openings among the trees; and as they passed along, they were deprived of this grand scenery only by another, and no less gratification that of finding themselves gliding through the deepest recesses of the forest, where nothing met the eye but the elegant windings of the clear and still Canal, and its borders adorned by a profusion of trees. . . . When the Proprietors approached the basin at Croydon, they saw it surrounded by many thousands of persons, assembled to greet, with thanks and applause, those by whose patriotic

perseverance so important a work had been accomplished. It is impossible to describe, adequately, the scene which presented itself, and the feelings which prevailed, when the Proprietors' barge was entering the basin, at which instant the band was playing 'God Save the King', the guns were firing, the bells of the churches were ringing; and this immense concourse of delighted persons were hailing by universal and hearty, and long continued shouts, the dawn of their commerce and prosperity.

The following air, written by a Gentleman, while sailing to Croydon, was most zealously and ably sung by one of the Proprietors, Mr J. Walsh, and was received with great applause:

All hail this grand day when with gay colours flying,
The barges are seen on the current to glide,
When with fond emulation all parties are vying,
To make our Canal of Old England the pride.

CHORUS
Long down its fair stream may the rich vessel glide,
And the Croydon Canal be of England the pride.

And may it long flourish, while commerce caressing,
Adorns its gay banks with her wealth-bringing stores;
To Croydon, and all round the country a blessing,
May industry's sons ever thrive on its shores!

And now my good fellows sure nothing is wanting
To heighten our mirth and our blessings to crown,
But with the gay belles on its banks to be flaunting,
When spring smiles again on this high-favoured town.[1]

Sometimes there were unofficial as well as official opening ceremonies, and these did not always work out quite so happily. When the lock joining the Gloucester and Berkeley Canal to the River Severn was opened, three local lads decided to brighten

[1] *The Times*, 27 October 1809.

up the proceedings by firing their own salute from a cannon – unfortunately, the cannon was ancient and the wadding was wet and the whole thing exploded as soon as it was lit. Mostly, though, there was no more serious accident to disturb the opening day than the occasional drunk falling into the canal.

Once the canal was open, there were even more interesting opportunities for the idly curious. There was, for example, the opportunity to take a trip on the canal in a pleasure boat. Canals-for-pleasure is no modern concept: Bentley, in an early version of his pamphlet promoting the Trent and Mersey Canal, suggested that gondolas might be introduced on to the canal – a proposal that conjures up romantic visions of England as a kind of overgrown Venice. Even if gondolas never did come to grace the Potteries, and there were no dark-eyed gondoliers to serenade the maidens of Stoke-on-Trent, the reality was romantic enough. The Reverend Shaw took a trip through the old Harecastle tunnel:

> I visited this tunnel about the year 1770, soon after it was finished, when pleasure boats were then kept for the purpose of exhibiting this great wonder; the impression it made on my mind, is still very fresh. The procession was solemn; some enlivened this scene with a band of musick, but we had none; as we entered far, the light of candles was necessary, and about half-way, the view back upon the mouth, was like the glimmering of a star, very beautiful.[1]

But the idyllic mood soon faded. The Trent and Mersey was built for commerce, not for sightseeing, and the sound of commerce would have soon drowned out the noise of the 'band of musick', even if one had come along.

> The various voices of the workmen from the mines, &c. were rude and aweful, and to be present at their quarrels, which sometimes happen when they meet, and battle for a passage, must resemble greatly the ideas we may form of the regions of Pluto.

[1] Rev. S. Shaw, *op. cit.*

It was all too much for the passenger. The mood was gone, and he soon gave up admiring the beauty and got down to thinking about the economic value of the tunnel and the canal in general:

> And though the expence attending this astonishing work was enormous, so as to promise little or no profit to the adventurers; yet in a few years after it was finished, I saw the smile of hope brighten every countenance, the value of manufactures arise in the most unthought of places; new buildings and new streets spring up in many parts of Staffordshire, where it passes; the poor no longer starving on the bread of poverty; and the rich grow greatly richer. The market town of Stone in particular soon felt this comfortable change; which from a poor insignificant place is now grown neat and handsome in its buildings, and from its wharfs and busy traffic, wears the lively aspect of a little seaport.

This was, of course, what the canals were really all about. For every spectator who came to satisfy his curiosity for seeing something new, there were half a dozen who came to see how prices were affected in towns along the route, the effects on manufacturers' profits and on the supply of food and other goods, and the income from tolls. The statistics for cost savings made impressive reading:

> The following statement of the differences of prices of carriage of goods, per ton, by Canal and Land, will strongly evince the great National Utility of Inland Navigation.[1]

	Canal	Land
Between Gainsbro and Birmingham	£1 10	£3 18
——— Manchester & Etruria, the centre of the Potteries	0 15	2 15
——— Ditto and Birmingham	1 10	4 0
——— Ditto and Stourport	1 10	4 13
——— Liverpool and Wolverhampton	1 5	5 0
——— Ditto and Birmingham	1 10	5 0

[1] *Felix Farley's Bristol Journal*, 1 December 1792.

	Canal	Land
——— Ditto and Stourport	1 10	5 0
——— Chester and Wolverhampton	1 15	3 10
——— Ditto and Birmingham	2 0	3 10
——— Ditto and Stourport	2 0	3 10

Thomas Pennant found that of all the sights that he saw on his tour through England, there was nothing that could inspire him to greater heights of enthusiasm than the contemplation of the economic benefits brought by the Trent and Mersey Canal:

> Notwithstanding the clamours which have been raised against this undertaking, in the places where it was intended to pass, when it was first projected, we have the pleasure now to see content reign universally on its banks, and plenty attend its progress. The cottage, instead of being half-covered with miserable thatch, is now secured with a substantial covering of tiles or slates, brought from the distant hills of Wales or Cumberland. The fields, which before were barren, are now drained and, by the assistance of manure, conveyed on the canal toll-free, are clothed with a beautiful verdure. Places which rarely knew the use of coal, are plentifully supplied with that essential article upon reasonable terms: and, what is still of greater public utility, the monopolizers of corn are prevented from exercizing their infamous trade; for, the communication being opened between Liverpool, Bristol and Hull, and the line of the canal being through countries abundant in grain, it affords a conveyance of corn unknown to past ages. At present, nothing but a general dearth can create a scarcity in any part adjacent to this extensive work.
>
> These, and many other advantages, are derived, both to individuals and the public, from this internal navigation. But when it happens that the kingdom is engaged in a foreign war, with what security is the trade between those three great ports carried on; and with how much less expence has the trader his goods conveyed to any part of the kingdom, than

he had formally been subject to, when the goods were obliged to be carried coastways, and to pay insurance?

I believe that it may be asserted, that no undertaking, equally expencive and arduous, was ever attempted by private people in any kingdom; and, in justice to the adventurers, it must be allowed, that, considering the difficulties they met with, owing to the nature of the works, or the caprice of persons whose lands were taken to make the canal, that ten years and a half was but a short time to perform it in; and that satisfaction has been made to every individual who suffered any injury by the execution of the undertaking. The profits arising from tonnage is already very considerable; and there is no doubt but that they will increase annually; and notwithstanding the enormous sum of money it has cost in the execution, the proprietors will be amply repaid, and have the comfort to reflect, that, by the conclusion of this project, they have contributed to the good of their country, and acquired wealth for themselves and posterity.[1]

All was for the best in the best of all possible worlds, served by inland navigation. Fifty years later, the Trent and Mersey still prospered, with quarter-shares, originally costing £50, being offered for £620. But the Panglossian dream of unending prosperity for all canals which was so vigorously propagated by Pennant and other writers of the eighteenth century failed to materialize. It scarcely had the opportunity. The gongoozlers of the nineteenth century found a new object of curiosity – the steam engine that could move along on rails and could pull wagons filled with goods and passengers. The canal builders had hardly finished their work before the new age was upon them. The Duke of Bridgewater's 'damned tramroads' came after all to put the canal system in the shade. Steadily, throughout the nineteenth century and into modern times, traffic on the canals declined. Some routes were kept open, many closed. Most of the existing waterways came under the control of the new railway companies, who had little incentive to maintain, let alone

[1] Thomas Pennant, *op. cit.*

improve, them. The process was inevitable. The canal system was created to meet the rapidly expanding transport needs of the industrial revolution, and it is no exaggeration to say that without that system the revolution could not have happened when it did, nor could it have spread in the way that it did. But the canal builders were also constructing the means to their own end: the waterways they built enabled industry to grow and flourish; that industrial growth in its turn fostered technological advances; and from those advances the steam engine developed as naturally as a plant from its seed. The development of the steam engine from its original static to its new mobile form came at just the right time. Once the first successful experiments with the locomotive had been made, there were no technical obstacles to putting the experiments to use. The railway engineers found a whole system of construction techniques and a body of skilled workmen ready and waiting for them – all by courtesy of the canal builders.

The canals were built to serve a commercial need. Tens of thousands of men were employed in their making and, for a brief period, they successfully met that need. The canals no longer fulfil that function, but they have found a new purpose, and every year now sees more and more pleasure boats thronging the waterway system. A new generation has discovered the delights of canal gongoozling. The canals have completed the transition from the practical to the picturesque, following the path described by Robert Southey for the Inverary Canal:

> The canal is a losing concern to the subscribers; and Mr Haddon complains that it draws off the water from the Don, to the hurt of his mills. . . . It is however a great benefit to the country, and no small ornament to it, with its clear water, its banks which are now clothed with weedery, and its numerous locks and bridges, all picturesque objects and pleasing, where you will find little else to look at.[1]

The whole of the modern development of canals for pleasure

[1] Robert Southey, *op. cit.*

boating, on a commercial scale, was foreseen over a century ago. Mr Robins, author of a *History of Paddington*, published in 1853, wrote:

> The glory of the first public company which shed its influence over Paddington has, in a great measure, departed; the shares of the Grand Junction Canal Company are below par, though the traffic on this silent highway is still considerable; and the cheap trips into the country offered by its means during the summer months are beginning to be highly appreciated by the people, who are pent in close lanes and alleys; and I have no doubt that shareholders' dividends would not be diminished by a more liberal attention to their want.

Mr Robins was a good prophet and, no doubt, he would be gratified to see his prophecy fulfilled. The canals have become accepted as a part of the countryside, and seem to fit in as naturally as the streams and rivers. But what are now ideal escape routes from the pressures of a modern industrial world were once the lifelines of the earliest industrial society. They cost the effort of an army of men; they involved individuals from all levels of society; they made some men rich, and brought bankruptcy to others. But the canal builders achieved one unique distinction – they left behind a transport system that enhances the countryside through which it passes instead of destroying it. That part, at least, of the builders' dream has come true.

The very last word ought to belong to the eighteenth century, the canal builders' own age. And the best person to pronounce it is the ballad writer, whose enthusiasm and *élan* might often have outstripped his abilities, but in whose verses we can still capture the feeling of the time and the excitement and the great promise that the coming of the canals brought to the country:

> Come now begin delving, the Bill is obtain'd
> The contest was hard, but a conquest is gain'd;
> Let no time be lost, and to get business done
> Set thousands to work, that work down the sun,

With speed the desirable work to compleat,
The hope how alluring – the spirit how great?
By Severn we soon, I've no doubt to my mind,
With old father Thames shall an intercourse find.

With pearmains and pippins 'twill gladden the throng,
Full loaded the boats to see floating along;
And fruit that is fine, and good hops for our ale,
Like Wednesbury pit-coal, will always find sale. . . .

As freedom I prize, and my Country respect,
I trust not a soul to my toast will object;
'Success to the Plough, not forgetting the Spade,
Health, plenty and peace, Navigation and Trade.'

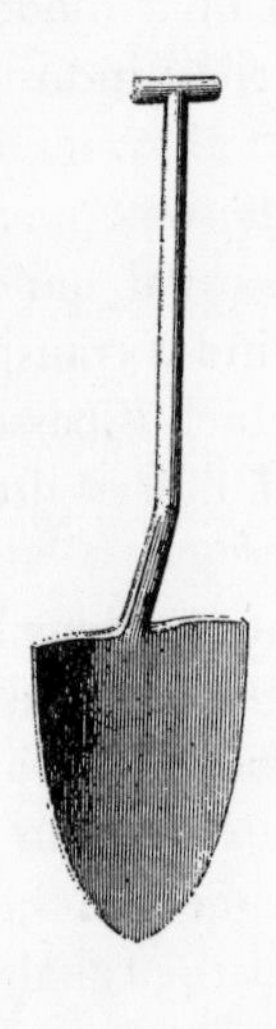

CHRONOLOGY

This chronology gives the dates for some of the more important events in the history of canal construction and development. Not all the Acts of Parliament listed as having been passed resulted in canals actually being built. The chronology is adapted from H. R. De Salis, *A Chronology of Inland Navigation in Great Britain* (1897) and Joseph Priestley, *Historical Account of the Navigable Rivers, Canals and Railways of Great Britain* (1831).

1759 Acts: First Bridgewater Act passed.
1760 Acts: Amended Bridgewater (including provision for the aqueduct at Barton).
1761 Construction: Bridgewater completed.
1762 Acts: Bridgewater extension to Runcorn.
1763–5 ———
1766 Acts: Trent and Mersey; Staffordshire and Worcestershire.
1767 ———
1768 Acts: Coventry; Droitwich; Birmingham (Birmingham to Autherley); Forth and Clyde; Borrowstowness.
1769 Acts: Oxford.
Construction: Birmingham canal completed from Wednesbury to Birmingham.
1770 Acts: Leeds and Liverpool; Monkland.
1771 Acts: Chesterfield; Bradford.
1772 Acts: Chester; Market Weighton.
Construction: Staffordshire and Worcestershire completed; Bridgewater joined to Mersey.
1773 ———
1774 Acts: Huddersfield Broad (Sir John Ramsden's); Bude.
Construction: Bradford completed.

1775 Acts: Gresley.

1776 Acts: Stourbridge; Dudley.
Construction: Chesterfield completed.

1777 Acts: Erewash.
Construction: Trent and Mersey completed; work stopped on Leeds and Liverpool (Leeds and Liverpool ends were both open).

1778 Acts: Basingstoke.
Construction: work stopped on Oxford (open to Banbury).

1779 Construction: Stroudwater open.

1780 Acts: Thames and Severn.

1781–2 ——————

1783 Acts: Birmingham and Fazeley.

1784–5 ——————

1786 Construction: work restarted on Oxford.

1787 ——————

1788 Acts: Shropshire (Tub Boat) Canal.
Construction: First canal inclined plane built by William Reynolds on Ketley Canal.

1789 Acts: Andover; Cromford.
Construction: Thames and Severn completed.

1790 Acts: Glamorganshire.
Construction: Oxford; Birmingham and Fazeley; Forth and Clye completed; work restarted on Leeds and Liverpool.

1791 Acts: Royal Military; Worcester and Birmingham; Hereford and Gloucester; Manchester, Bolton and Bury; Kington and Leominster; Neath.

1792 Acts: Nottingham; Ashton; Lancaster; Wyrley and Essington; Coombe Hill; Monmouthshire.
Construction: Shropshire completed.

1793 Acts: Oakham; Grantham; Ulverston; Nutbrook; Derby; Grand Junction; Caistor; Shrewsbury; Stainforth and Keadby; Dearne and Dove; Stratford-upon-Avon; Brecnock and Abergavenny; Ellesmere; Warwick and Birmingham; Old Union; Gloucester and Berkeley; Aberdare; Barnsley; Crinan.

1794 Acts: Montgomeryshire; Somersetshire Coal Canal; Wisbech; Peak Forest; Huddersfield Narrow; Swansea; Grand Junction (branches to Buckingham, Aylesbury and Wendover); Warwick and Braunston; Rochdale; Kennet and Avon; Ashby.
Construction: Glamorganshire Canal opened from Merthyr Tydfil to Cardiff.

1795 Acts: Bridgewater (extension to Leigh); Grand Junction (Paddington branch; Wiltshire and Berkshire; Newcastle under Lyme; Ivelchester and Langport.

1796 Acts: Salisbury and Southampton; Grand Western; Warwick and Napton; Aberdeenshire.
Construction: Lune aqueduct completed.

1797 Acts: Polbrock.
Construction: Ashton; Shrewsbury completed.

1798 Construction: Swansea completed; Huddersfield Narrow (Huddersfield to Marsden and Ashton to Staleybridge) opened; Hereford and Gloucester (Gloucester to Ledbury) opened.

1799 Construction: Warwick and Birmingham; Warwick and Napton; Barnsley completed.

1800 Acts: Thames and Medway.
Construction: Peak Forest completed.

1801 Acts: Croydon; Grand Surrey; Leven.
Construction: Grand Junction (Buckingham and Paddington branches) completed.

1802 Construction: Nottingham completed.

1803 Acts: Tavistock; Caledonian.
Construction: Pontcysyllte aqueduct completed.

1804 Construction: Dearne and Dove; Rochdale completed.

1805 Construction: Grand Junction (main line from Brentford to Braunston) completed.

1806 Acts: Glasgow, Paisley and Johnstone.

1807 Acts: Isle of Dogs.

1808 ————————

1809 Construction: Perpendicular lift, invented by John Woodhouse, erected at Tardebigge on Worcester and Birmingham.

1810 Acts: Grand Union.
Construction: Kennet and Avon completed.

1811 Acts: Bridgewater and Taunton.

1812 Acts: Regent's; North Walsham and Dilham; London and Cambridge Junction.

1813 Acts: Wey and Arun Junction; North Wilts.
Construction: Grand Junction (Aylesbury branch) completed.

1814 Acts: Newport Pagnell.
Construction: Grand Western (Loudwell to Tiverton) completed.

1815 Acts: Pocklington; Sheffield.
Construction: Grand Junction (Northampton branch) completed.

1816 Construction: Leeds and Liverpool completed.

1817 Acts: Portsmouth and Arundel; Edinburgh and Glasgow Union.

1818 ———

1819 Acts: Bude; Carlisle.
Construction: North Wiltshire completed.

1820 Construction: Regent's completed.

1821 Act passed for Stockton and Darlington Railway.

1822 Construction: Caledonian; Edinburgh and Glasgow Union completed.

1823 Acts: Harecastle New Tunnel.

1824 Acts: Kensington; Hertford Union.

1825 Acts: Baybridge; Liskeard; English and Bristol Channels Ship Canal.

1826 Acts: Birmingham and Liverpool Junction; Macclesfield; Alford.
Construction: Lancaster completed.

1827 Construction: Harecastle New Tunnel; Gloucester and Berkeley completed.

BIBLIOGRAPHY

The majority of the eighteenth- and nineteenth-century sources have already been listed in the footnotes to the text. Almost all the canal company records are to be found in the British Transport Historical Records, 60 Porchester Road, London W2. A few other sources are worth special mention:

CARY, J., *Inland Navigation* (1795), contains details and maps of canals.

DE SALIS, H. R., *A Handbook of Inland Navigation* (1901), is the Bradshaw of the canals. It contains a complete gazetteer with details of routes, levels and distances, and also contains a brief but useful technical essay and glossary.

———, *Chronology of Inland Navigation* (1897), is what its name suggests – a year-by-year listing of the most important events in canal history.

HASSELL, J., *Tour of the Grand Junction* (1812), contains an interesting eye-witness account but is chiefly notable for Hassell's illustrations.

PHILLIPS, J., *A General History of Inland Navigation* (1795), has already been quoted at length in the text, and contains one of the best available accounts of the early canals.

PLYMLEY, J., *A General View of the Agriculture of Shropshire* (1813), contains an article on canals by Thomas Telford.

PRIESTLEY, J., *Historical Account of the Navigable Rivers, Canals and Railways of Great Britain* (1831), is one of the most useful of the early publications, and contains details of the various Acts of Parliament affecting canals and navigation.

SMILES, S., *Lives of the Engineers* (5 vols, 1862), contains biographies of many of the men connected with canal building.

Travellers' accounts of canals can be found in two collections:

MAVOR, W., *The British Tourists* (6 vols, 1809).

PINKERTON, J., *A General Collection of the Best and Most Interesting Voyages* (Vols 1–3, 1808).

There has been a great increase in the production of canal histories in recent years. The most notable example is the series 'The Canals of the British Isles', edited by Charles Hadfield. The first volume is a general survey, and subsequent volumes trace the history of the various canals and tramways of the region from their inception up to the present day:

HADFIELD, C., *British Canals: An Illustrated History* (3rd edition, 1968).

———, *The Canals of the East Midlands* (1966).

———, *The Canals of South Wales and the Border* (1960).

———, *The Canals of Southern England* (1955).

———, *The Canals of the West Midlands* (1966).

LINDSAY, J., *The Canals of Scotland* (1968).

Other general histories of canals include:

DE MARÉ, E., *The Canals of England* (1951), is chiefly notable for Eric de Maré's superb photopraphs.

HADFIELD, C., *The Canal Age* (1969), is an account of canals all over the world.

ROLT, L. T. C., *Navigable Waterways* (1969), is a volume in the 'Industrial Archaeology' series, edited by Rolt, and deals with the technology of the canals and the canal system as it is today.

——— *Inland Waterways of England* (1950), is one of the standard works on the history of navigations and canals.

SMITH, P., *Waterways Heritage* (1971), is a compilation of early illustrations and facsimiles of documents.

A number of books have been published dealing with specific waterways:

CLEW, K. R., *The Kennet and Avon Canal* (1968).

EWANS, M. C., *The Haytor granite tramway and Stover Canal* (1964).

HADFIELD, C., and NORRIS, J., *Waterways to Stratford* (1962).

TEW, D., *The Oakham Canal* (1968).

VINE, P. A. L., *London's Lost Route to the Sea: an historical account of the Inland Navigations which linked the Thames to the English Channel* (1965).

A general history of tramways is available:

BAXTER, B., *Stone Blocks and Iron Rails* (1966).

Finally, there are a few modern biographies of canal engineers:

BOUCHER, C., *John Rennie* (1963).

MALET, H., *The Canal Duke* (1961).

ROLT, L. T. C., *Thomas Telford* (1958).

INDEX